Mathematics Tomorrow

WITHDRAWN FROM STOCK

Mathematics Tomorrow

Edited by
Lynn Arthur Steen

With 14 Illustrations

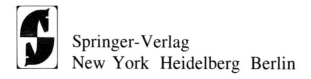
Springer-Verlag
New York Heidelberg Berlin

Dr. Lynn Arthur Steen
Saint Olaf College
Northfield, Minnesota 55057
USA

Library of Congress Cataloging in Publication Data
Main entry under title:
Mathematics tomorrow.
    1. Mathematics—Addresses, essays, lectures.
I. Steen, Lynn Arthur, 1941—. II. Halmos,
Paul Richard, 1914—.
QA7.M3448   510   81-370
                        AACR1

© 1981 by Springer-Verlag New York Inc.
All rights reserved. No part of this book may be translated or reproduced
in any form without written permission from Springer-Verlag, 175 Fifth
Avenue, New York, New York 10010, USA.
The use of general descriptive names, trade names, trademarks, etc. in this
publication, even if the former are not especially identified, is not to
be taken as a sign that such names, as understood by the Trade Marks and
Merchandise Marks Act, may accordingly be used freely by anyone.

Printed in the United States of America.

9 8 7 6 5 4 3 2 1

ISBN 0-387-**90564-2** Springer-Verlag New York Heidelberg Berlin
ISBN 3-540-**90564-2** Springer-Verlag Berlin Heidelberg New York

Contents

Issues of Equality

Mathematics for Tomorrow

Introduction

Mathematics today is approaching a state of crisis. As the demands of science and society for mathematical literacy increase, the percentage of American college students intending to major in mathematics plummets and achievement scores of entering college students continue their unremitting decline. As research in core mathematics reaches unprecedented heights of power and sophistication, the growth of diverse applied specialties threatens to fragment mathematics into distinct and frequently hostile mathematical sciences.

These crises in mathematics presage difficulties for science and engineering, and alarms are beginning to sound in the scientific and even in the political communities. Citing a trend towards "virtual scientific and technological illiteracy" and a "shrinking of our national commitment to excellence ... in science, mathematics and technology," a recent study conducted for the President by the U.S. National Science Foundation and Department of Education warns of serious impending shortcomings in public understanding of science. "Today people in a wide range of non-scientific ... professions must have a greater understanding of technology than at any time in our history. Yet our educational system does not now provide such understanding." The study goes on to conclude that present trends pose great risk of manpower shortages in the mathematical and engineering sciences. "The pool from which our future scientific and engineering personnel can be drawn is ... in danger of becoming smaller, even as the need for such personnel is increasing." It is time to take a serious look at mathematics tomorrow.

This volume of essays examining the possible shape of mathematics in the future is a sequel to the volume *Mathematics Today: Twelve Informal Essays* prepared three years ago by the Joint Projects Committee on Mathematics and the Conference Board of Mathematical Sciences, with the assistance of a grant from the National Science Foundation. Although *Mathematics Tomorrow* continues the theme of *Mathematics Today*, it differs from it in many respects. Most important, *Mathematics Tomorrow* is a collection of opinions and predictions about the direction that mathematics research and mathematics education should take in the immediate future. It is not a project of the Conference Board or of the Joint Project Committee on Mathematics; neither of these organizations is in any way responsible for the opinions expressed in this volume. *Mathematics Tomorrow* is the product of many individuals, not the work of any official mathematics organization.

The mathematicians who have contributed to this volume are experienced teachers and scholars who represent diverse elements of the mathematical community. They speak here as individuals deeply concerned about the direction of mathematics and mathematics education. Some argue for radical reform, citing as evidence the sudden explosion in demand for mathematical modelling and discrete, computer-based mathematics. Others stress the importance of conserving the traditional value of mathematics as the formal expression of pure structure that provides science with both necessary logic and valuable metaphor. All are concerned with the manner in which we transmit the nature and value of mathematics to our children. Mathematics tomorrow is the mathematics education of today.

We begin this volume with an inquiry into the nature of mathematics that reflects an ancient counterpoint between pure and applied mathematics. This counterpoint is a vivid expression of the dual character of mathematics. Mathematics is both elegant and powerful; its standards of truth are both aesthetic and pragmatic; it is truly both an art and a science. Yet many who study or practice mathematics feel the influence of these features unequally.

In our first section Paul Halmos, Jerry King and Tim Poston restate—in quite different ways—the importance of enduring truth and abstract beauty. Mathematics reveals for Halmos the "breaktakingly complicated" logical structure of the universe; for King, an "aesthetic value" as clearly defined as that of music or poetry; and for Poston, an "intellectual harmony" as smooth and clear as a musical tone. Mathematics, for these authors, differs from applied mathematics as a poem differs from legislative prose.

Jerome Spanier, on the other hand, argues that the power of mathematics is not fully revealed by its internal structure, that "solving equations" is not the same as solving problems. Mathematics for Spanier is not an isolated intellectual structure but a part of the general process of scientific modelling. According to Alan Tucker, today's pragmatic students are, in fact, "voting with their feet," drawing their main intellectual sustenance from the vigorous new applied areas of the mathematical sciences. Indeed, contemporary science provides a regular supply of new problems and fruitful structures for modern mathematics. In the concluding essay of this first section, William Lucas explores the new components of this growth, arguing that decision science, for example, is now posing for mathematics challenges akin to those posed by physics a generation ago. The nature of mathematics continues to be profoundly affected by the kind of problems it receives from science.

That tension exists between pure and applied mathematics is neither newsworthy nor regrettable. Indeed, this tension is a major source of new mathematics, as theory first catches up with practice, then practice with new theory. The pattern is as old as mathematics itself. The Greek study of the pure forms of conic sections led, some two thousand years later, to scientific models of planetary orbits, while the business arithmetic of the ancient Egyptians has led in modern time to esoteric results in what G.H. Hardy boastingly called the most useless of all mathematics—the theory of numbers. These ancient examples illustrate the compelling power of both pure and applied mathematics as well as the bonds that link them. Similar evidence abounds in today's mathematics: recently created abstract theories of graphs and matrices make possible the design of computer systems that are revolutionizing the way we live and work, while the computer revolution itself is generating whole new fields of "pure" mathematical research, in, for example, data structures and analysis of algorithms.

What is newsworthy about the explosion of mathematical theory and its applications is that so many citizens are either oblivious to it or afraid of it. As science and society come to depend more and more on the fruits of mathematics, we hear more and more about mathematical anxieties, phobias and illiteracy. As demands for mathematical education increase, both achievements and attitudes seem to worsen. Indeed, the Presidental study cited above reports that only one-third of U. S. school districts require graduates to take more than one year of mathematics. "More students than ever before are dropping out of science and mathematics courses after the tenth grade, and this trend shows no signs of abating." The real crisis of mathematics today is that mathematics education is so feeble. It is hard to see, under these circumstances, how mathematics itself can remain strong.

Peter Hilton argues, in his essay opening our section on teaching and learning mathematics, that math avoidance is no pathology: it is, he argues, a "thoroughly healthy" reaction to a traditional elementary school mathematics curriculum which is, all too often, deplorable, pointless and dull. Today's teachers were yesterday's pupils, as today's pupils will become tomorrow's teachers. So anxiety breeds anxiety, in an unending cycle. "The more anxious a teacher is," report Anneli Lax and Giuliana Groat in the second essay of this section, "the more he retreats into the trodden path of memorized rules and rote instruction." Lax and Groat go on to urge that mathematics instruction should cultivate a vigorous interaction of analytic and heuristic thinking. Continuing this theme, Abe Shenitzer, in the next essay, provides numerous examples of how mathematics can be taught heuristically and genetically by embedding topics in appropriate contexts. Harold Edwards in "Read the Masters!" suggests that history is an important and frequently overlooked context. "History has shaped what we are, and it is our best source of information as to what is possible."

Neal Koblitz, in contrast, illustrates the power of mathematics when used in the wrong context: he shows how mathematics can, by means of propagandists' cleverly contrived examples, easily mystify rather than clarify, intimidate rather than help, or "convey a false impression of precision and profundity." Koblitz' essay suggests numerous issues for mathematics education, not least the goal of educating citizens to the point where they cannot be easily conned by mathematical or statistical sleight of hand.

The final four essays in this section deal more with the context of mathematics education than with its content. Three experienced mathematics publishers—Walter Kaufmann-Bühler and Klaus and Alice Peters—set forth speculative yet serious ideas on how the role of books and publishers may change in an era of electronic editing and home computers. Donald Albers and George Miller report on the prospects for mathematics education in the two year colleges, which now enroll slightly more than half of all students in higher education. The problems of faculty in these schools are serious. About half of the mathematics teachers are part-time, very few are active in curriculum development or professional growth, and the majority are entering middle age unprepared for the revolution in mathematical science that their students are already experiencing.

It is clear from all accounts that mathematics teachers will require substantial assistance if they are to meet the special challenge of preparing students for the new mathematics of the 1980's. E. P. Miles recounts, in the final essay of the second section, how the U.S. National Science Foundation has waxed and waned in support of teacher training and curriculum development. At its peak—in the post-Sputnik period of the early '60's—NSF funds were divided about equally between research and education. Now science education accounts for less than 10% of the NSF budget, a reflection perhaps of a long national inattention that has also contributed to the phobias, anxieties and underachievement that now seem endemic in contemporary U.S. mathematics education. Comparison of U.S. mathematics educational achievement with that in the U.S.S.R. (in a recent report by Isaac Wirzup to the National Science Foundation) reveals a ratio between the Soviet Union and the United States of approximately 20:1 in the number of youths reaching general scientific literacy in calculus, physics and chemistry. The Presidential study (cited above) underscores this concern: "The declining emphasis on science and mathematics in our school systems is in marked contrast to other industrialized countries. Japan, Germany, and the Soviet Union all provide rigorous training in science and mathematics for all their citizens."

The impending U.S. manpower shortages in computer science and engineering are likely to produce some change in the pattern of support for mathematics education in the years ahead. But even these possible changes are unlikely to reverse the current flight of teachers from the school classroom. In 1980-81 almost 25% of the mathematics teaching positions in the U.S. were filled by teachers without appropriate certification or credentials. This shortage of qualified secondary school mathematics teachers is likely to become dramatically more serious in the 1980's, as both college graduates and experienced teachers migrate to higher paying jobs in the computer and high technology industries.

The third section of our look at mathematics tomorrow examines from three perspectives one of our age's most vexing problems—the alarming drop-out rate of young girls from school science and mathematics courses. Eileen Poiani and Alice Schafer examine the role played by mathematics as a "critical filter" in career choices. "Like it or not, mathematics opens career doors;" yet women average 50 points behind men on mathematics SAT scores, and the ratio of women to men in courses such as calculus is only about 1:4 and rising very slowly. To offer a context for our present concerns, Schafer recounts in her essay the impediments and achievements of six distinguished women mathematicians: some were denied enrollment at university, some were enrolled but then denied their degrees, and all were denied jobs for which they were well qualified. Marian Pour-El offers a modern counterpoint by her revealing personal account of family and career patterns facing a contemporary woman mathematician or scientist. As these essays strongly suggest, one answer to our current shortage of mathematically trained personnel is to utilize more fully the scientific and mathematical talents of women.

In our final section we offer various glimpses of mathematics tomorrow, suggestions of how mathematics may evolve into a loosely linked cluster of diverse mathematical sciences. Several forces have united in recent years to thrust mathematics deeply into various applied specialties: the rise of computing, the pragmatism of contemporary students, the demands of newly quantified sciences, and public reaction to what were perceived as excesses and failures of the theoretically inclined "new math." For these and other reasons, applications have become the theme of virtually all contemporary curriculum reform.

The first essay in our final section, by Ross Finney, provides simple yet effective examples of applied mathematics from some innovative, modular curricular materials developed by the Undergraduate Mathematics Applications Project (UMAP). Finney's examples show not just that mathematics can be used in interesting ways—he cites determination of blood flow, scheduling of prison guards, analysis of elected legislative bodies—nor just that mathematics teaching can be made more appealing by use of such examples, but that real applied mathematics relies on strategies and decisions rarely found in typical textbook problems.

The next four essays discuss in some detail four specific areas of applied mathematics: Anthony Ralston introduces the major new area of discrete, combinatorial mathematics, a field that has blossomed in today's computer environment. Ralston suggests that discrete mathematics is the paradigm of our age, as calculus was for former ages. Following this, Paul Boggs discusses the unique role of mathematical software as a product that mathematics can exhibit and even sell: software is an important symbiosis of mathematics and computer science that is economically valuable, easily demonstrable, and almost tangible.

Hartley Rogers, Tim Robertson and Robert Hogg discuss two of mathematics' close relatives: physics and statistics. Each of these fields contributes, albeit in quite different ways, to the development of valuable mathematical insight. The evolving relations between these fields and mathematics will do much to shape the nature of mathematics education in the coming decade. Finally, Maynard Thompson examines the general relation between mathematics and science, focusing especially on the differences between mathematical modelling in the mature physical sciences and modelling in the newly emerging life and social sciences.

The view of mathematics provided by these essays is like the beginning of an explosion in slow motion. Powerful forces, unleashed by the pure intellectual energy of core mathematics, are pushing mathematical ideas in every direction. As the waves of new ideas pass through adjacent scientific fields, they are amplified and reshaped to fit new bodies of knowledge and new methods of inquiry. Although eventually these new forces will become part of science, they are now a moving, churning, tumultuous part of applied mathematics. The energy in the core of mathematics, however, is as great as ever, although fewer people these days penetrate to the core. Those who do report powerful new bonds that unify the most central concepts of core mathematics. This new unity—discussed more in *Mathematics Today* than in the present volume—insures an indefinite continuation of the intellectual force that is powering the present explosion of applied mathematics. The challenge for mathematics tomorrow is to harness this force and insure that the next generation can use it effectively.

Lynn Arthur Steen
Northfield, Minnesota
January, 1981

What Is Mathematics?

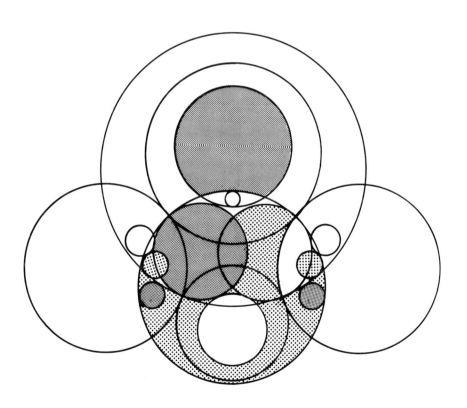

Applied Mathematics Is Bad Mathematics

Paul R. Halmos

It isn't really (applied mathematics, that is, isn't really bad mathematics), but it's different.

Does that sound as if I had set out to capture your attention, and, having succeeded, decided forthwith to back down and become conciliatory? Nothing of the sort! The "conciliatory" sentence is controversial, believe it or not; lots of people argue, vehemently, that it (meaning applied mathematics) is not different at all, it's all the same as pure mathematics, and anybody who says otherwise is probably a reactionary establishmentarian and certainly wrong.

If you're not a professional mathematician, you may be astonished to learn that (according to some people) there are different kinds of mathematics, and that there is anything in the subject for anyone to get excited about. There are; and there is; and what follows is a fragment of what might be called the pertinent sociology of mathematics: what's the difference between pure and applied, how do mathematicians feel about the rift, and what's likely to happen to it in the centuries to come?

What is it?

There is never any doubt about what mathematics encompasses and what it does not, but it is not easy to find words that describe precisely what it is. In many discussions, moreover, mathematics is not described as a whole

Paul R. Halmos is Distinguished Professor of Mathematics at Indiana University, and Editor-Elect of the *American Mathematical Monthly*. He received his Ph.D. from the University of Illinois, and has held positions at Illinois, Syracuse, Chicago, Michigan, Hawaii, and Santa Barbara. He has published numerous books and nearly 100 articles, and has been the editor of many journals and several book series. The Mathematical Association of America has given him the Chauvenet Prize and (twice) the Lester Ford award for mathematical exposition. His main mathematical interests are in measure and ergodic theory, algebraic logic, and operators on Hilbert space.

but is divided into two parts, and not just in one way; there are two kinds of mathematics according to each of several different systems of classification.

Some of the dichotomies are well known, and others less so. Mathematics studies sizes and shapes, or, in other words, numbers (arithmetic) and figures (geometry); it can be discrete or continuous; it is sometimes finite and sometimes infinite; and, most acrimoniously, some of it is pure (useless?) and some applied (practical?). Different as these classification schemes might be, they are not unrelated. They are, however, not of equal strengths; the size-shape division, for instance, is much less clear-cut, and much less divisive, than the pure-applied one.

Nobody is forced to decide between vanilla ice cream and chocolate once and for all, and it is even possible to mix the two, but most people usually ask for the same one. A similar (congenital?) division of taste exists for mathematicians. Nobody has to decide once and for all to like only algebra (discrete) or only topology (continuous), and there are even flourishing subjects called algebraic topology and topological algebra, but most mathematicians do in fact lean strongly toward either the discrete or the continuous.

Squares and spheres

It would be a shame to go on and on about mathematics and its parts without looking at a few good concrete examples, but genuine examples are much too technical to describe in the present context. Here are a couple of artificial ones (with some shortcomings, which I shall explain presently).

Suppose you want to pave the floor of a room whose shape is a perfect square with tiles that are themselves squares so that no two tiles are exactly the same size. Can it be done? In other words, can one cover a square with a finite number of non-overlapping smaller squares all of which have different side-lengths? This is not an easy question to answer.

Here is another puzzle: if you have a perfect sphere, like a basketball, what's the smallest number of points you can mark on it so that every point on the surface is within an inch of one of the marked ones? In other words, what's the most economical way to distribute television relay stations on the surface of the globe?

Is the square example about sizes (numbers) or shapes (figures)? The answer seems to be that it's about both, and so is the sphere example. In this respect the examples give a fair picture; mixed types are more likely to occur (and are always more interesting) than the ones at either extreme. The examples have different flavors, however. The square one is more nearly arithmetic, discrete, finite, pure, and the one about spheres leans toward being geometric, continuous, infinite, applied.

The square problem is of some mild interest, and it has received attention in the professional literature several times, but it doesn't really have the respect of most mathematicians. The reason is not that it is obviously useless in the practical sense of the word, but that it is much too special (petty?, trivial?), in the sense of being isolated from most of the rest of mathematics and requiring *ad hoc* methods for its solution. It is not really a fair example of pure mathematics.

The sphere example, on the other hand, is of a great deal of practical use, but, nevertheless, it is not a fair example of applied mathematics: it is much easier (and much purer) than most applied problems, and, in particular, it does not involve motion, which plays the central role in the classical conception of what applied mathematics is all about.

Still, for what they are worth, here they are, and it might help to keep them in mind as the discussion proceeds.

Fiction or action

The pure and applied distinction is visible in the arts and in the humanities almost as clearly as in the sciences: witness Mozart versus military marches, Rubens versus medical illustrations, or Virgil's *Aeneid* versus Cicero's *Philippics*. Pure literature deals with abstractions such as love and war, and it tells about imaginary examples of them in emotionally stirring language. Pure mathematics deals with abstractions such as the multiplication of numbers and the congruence of triangles, and it reasons about Platonically idealized examples of them with intellectually convincing logic.

There is, to be sure, one sense of the word in which all literature is "applied". Shakespeare's sonnets have to do with the everyday world, and so does Tolstoy's *War and Peace*, and so do Caesar's commentaries on the wars he fought; they all start from what human beings see and hear, and all speak of how human beings move and feel. In that same somewhat shallow sense all mathematics is applied. It all starts from sizes and shapes (whose study leads ultimately to algebra and geometry), and it reasons about how sizes and shapes change and interact (and such reasoning leads ultimately to the part of the subject that the professionals call analysis).

There can be no doubt that the fountainhead, the inspiration, of all literature is the physical and social universe we live in, and the same is true about mathematics. There is also no doubt that the physical and social universe daily affects each musician, and painter, and writer, and mathematician, and that therefore a part at least of the raw material of the artist is the world of facts and motions, sights and sounds. Continual contact between the world and art is bound to change the latter, and perhaps even to improve it.

The ultimate goal of "applied literature", and of applied mathematics, is action. A campaign speech is made so as to cause you to pull the third lever on a voting machine rather than the fourth. An aerodynamic equation is solved so as to cause a plane wing to lift its load fast enough to avoid complaints from the home owners near the airport. These examples are crude and obvious; there are subtler ones. If the uiography of a candidate, a factually correct and honest biography, does not directly mention the forthcoming election, is it then pure literature? If a discussion of how mathematically idealized air flows around moving figures of various shapes, a logically rigorous and correct discussion, does not mention airplanes or airports, is it then pure mathematics? And what about the in-between cases: the biography that, without telling lies, is heavily prejudiced, and the treatise on aerodynamics that, without being demonstrably incorrect, uses cost-cutting rough approximations—are they pure or applied?

Continuous spectrum

Where are the dividing lines in the chain from biography to interpretive history to legend to fiction? We might be able to tell which of Toynbee, Thucydides, Homer, and Joyce is pure and which is applied, but if we insert a dozen names between each pair of them, as we pass from inter-preted fact to pure fancy, the distinctions become blurred and perhaps impossible to define. The mathematical analogy is close: if we set out to sort a collection of articles that range from naval architecture to fluid dynamics to partial differential equations to topological vector spaces, the pure versus applied decisions that are clear at the two ends of the spectrum become fuzzy in the middle.

Pure mathematics can be practically useful and applied mathematics can be artistically elegant.

To confuse the issue still more, pure mathematics can be practically useful and applied mathematics can be artistically elegant. Pure mathematicians, trying to understand involved logical and geometrical interrelations, discovered the theory of convex sets and the algebraic and topological study of various classes of functions. Almost as if by luck, convexity has become the main tool in linear programming (an indispensable part of modern economic and industrial practice), and functional analysis has become the main tool in quantum theory and particle physics. The physicist regards the applicability of von Neumann algebras (a part of functional analysis) to elementary particles as the only justification of the former; the mathematician regards the connection as the only interesting aspect of the latter. *De gustibus non disputandum est?*

Just as pure mathematics can be useful, applied mathematics can be more beautifully useless than is sometimes recognized. Applied mathematics is not engineering; the applied mathematician does not design airplanes or atomic bombs. Applied mathematics is an intellectual discipline, not a part of industrial technology. The ultimate goal of applied mathematics is action, to be sure, but, before that, applied mathematics is a part of theoretical science concerned with the general principles behind what makes planes fly and bombs explode.

The differences between people are sometimes as hard to discern as the differences between subjects, and it can even happen that one and the same person is both a pure and an applied mathematician. Some applied mathematicians (especially the better ones) have a sound training in pure mathematics, and some pure mathematicians (especially the better ones) have a sound training in applicable techniques. When the occasion for a crossover arises (a pure mathematician successfully solves a special case of the travelling salesman problem that arises in operations research, a relativity theorist brilliantly derives a formula in 4-dimensional differential geometry), each one is secretly more than a little proud: "See! I can do that stuff too!"

Doers and knowers

What I have said so far is that in some sense all mathematics is applied mathematics and that on some level it is not easy to tell the pure from the applied. Now I'll talk about the other side: pure and applied are different indeed, and if you know what to look for, and have the courage of your convictions, you can always tell which is which. My purpose is to describe something much more than to prove something. My hope is neither to convert the pagan nor to convince the agnostic, but just to inform the traveller from another land: there are two sects here, and these are the things that they say about each other.

The difference of opinion is unlike the one in which one sect says "left" and the other says "right"; here one sect says "we are all one sect" and the other says "oh no, we're not, we are two". That kind of difference makes it hard to present the facts in an impartial manner; the mere recognition that a conflict exists amounts already to taking sides. There is no help for that, so I proceed, with my own conclusions admittedly firm, to do the best I can for the stranger in our midst.

Human beings want to know and to do. People want to know what their forefathers did and said, they want to know about animals and vegetables and minerals, and they want to know about concepts and numbers and sights and sounds. People want to grow food and to sew clothes, they want to build houses and to design machines, and they want to cure diseases and to speak languages.

The doers and the knowers frequently differ in motivation, attitude, technique, and satisfaction, and these differences are visible in the special case of applied mathematicians (doers) and pure mathematicians (knowers). The motivation of the applied mathematician is to understand the world and perhaps to change it; the requisite attitude (or, in any event, a customary one) is one of sharp focus (keep your eye on the problem); the techniques are chosen for and judged by their effectiveness (the end is what's important); and the satisfaction comes from the way the answer checks against reality and can be used to make predictions. The motivation of the pure mathematician is frequently just curiosity; the attitude is more that of a wide-angle lens than a telescopic one (is there a more interesting and perhaps deeper question nearby?); the choice of technique is dictated at least in part by its harmony with the context (half the fun is getting there); and the satisfaction comes from the way the answer illuminates unsuspected connections between ideas that had once seemed to be far apart.

The challenge [for the pure mathematician] is the breathtakingly complicated logical structure of the universe, and victory is permanent.

The last point deserves emphasis, especially if you belong to the large group of people who proudly dislike mathematics and regard it as inglorious drudgery. For the pure mathematician, his subject is an inexhaustible source of artistic pleasure: not only the excitement of the puzzle and the satisfaction of the victory (if it ever comes!), but mostly the joy of contemplation. The challenge doesn't come from our opponent who can win only if we lose, and victory doesn't disappear as soon as it's achieved (as in tennis, say); the challenge is the breathtakingly complicated logical structure of the universe, and victory is permanent (more like recovering precious metal from a sunken ship).

The basic differences in motivation, attitude, technique, and satisfaction are probably connected with more superficial but more noticeable differences in exposition. Pure and applied mathematicians have different traditions about clarity, elegance, and perhaps even logical rigor, and such differences frequently make for unhappy communication.

The hows and the whys listed above are not offered as a checklist to be used in distinguishing applied science from pure thought; that is usually done by a sort of intuitive absolute pitch. The word "spectrum" gives an analogical hint to the truth. In some sense red and orange are the same— just waves whose lengths differ a bit—and it is impossible to put your finger on the spot in the spectrum where red ends and orange begins—but, after that is granted, red and orange are still different, and the task of telling them apart is almost never a difficult one.

Beauty and boredom

Many pure mathematicians regard their specialty as an art, and one of their terms of highest praise for another's work is "beautiful". Applied mathematicians seem sometimes to regard their subject as a systematization of methods; a suitable word of praise for a piece of work might be "ingenious" or "powerful".

> *Mathematics, . . . despite its many subdivisions and their enormous rates of growth, . . . is an amazingly unified intellectual structure.*

Here is another thing that has frequently struck me: mathematics (pure mathematics), despite its many subdivisions and their enormous rate of growth (started millennia ago and greater today than ever before), is an amazingly unified intellectual structure. The mathematics that is alive and vigorous today has so many parts, and each is so extensive, that no one can possibly know them all. As a result, we, all of us, often attend colloquium lectures on subjects about which we know much less than an average historian, say, knows about linguistics. It doesn't matter, however, whether the talk is about unbounded operators, commutative groups, or parallelizable surfaces; the interplay between widely separated parts of mathematics always shows up. The concepts and methods of each one illuminate all others, and the unity of the structure as a whole is there to be marvelled at.

That unity, that common aesthetic insight, is mostly missing between pure and applied mathematics. When I try to listen to a lecture about fluid mechanics, I soon start wondering and puzzling at the (to me) *ad hoc* seeming approach; then the puzzlement is replaced by bewilderment, boredom, confusion, acute discomfort, and, before the end, complete chaos. Applied mathematicians listening to a lecture on algebraic geometry over fields of non-zero characteristic go through a very similar sequence of emotions, and they describe them by words such as inbred, artificial, baroque folderol, and unnecessary hairsplitting.

It might be argued that from the proper, Olympian, impartial scientific point of view both sides are wrong, but perhaps to a large extent both are right—which would go to prove that we are indeed looking at two subjects, not one. To many pure mathematicians applied mathematics is nothing but a bag of tricks, with no merit except that they work, and to many applied mathematicians much of pure mathematics deserves to be described as meaningless abstraction for its own sake with no merit at all. (I mention in passing that in moments of indulgent self-depreciation the students of one particular branch of pure mathematics, category theory, refer to their branch as "abstract nonsense"; applied mathematicians tend to refer to it the same way, and they seem to mean it.)

New heresy

Some say that the alleged schism between pure and applied mathematics is a recent heresy at which the founding greats would throw up their hands in horror—the world is going to the dogs! There is a pertinent quotation from Plato's *Philebus*, which doesn't quite refute that statement, but it's enough to make you think about it again.

"Socrates: Are there not two kinds of arithmetic, that of the people and that of the philosophers? . . . And how about the arts of reckoning and measuring as they are used in building and in trade when compared with philosophical geometry and elaborate computations—shall we speak of each of these as one or two?

"Protarchus: . . . I should say that each of them was two."

Is the distinction that Socrates is driving at exactly the pure-applied one? If not that, then what?

The only other curiosity along these lines that I'll mention is that you can usually (but not always) tell an applied mathematician from a pure one just by observing the temperature of his attitude toward the same-different debate. If he feels strongly and maintains that pure and applied are and must be the same, that they are both mathematics and the distinction is meaningless, then he is probably an applied mathematician. About this particular subject most pure mathematicians feel less heat and speak less polemically: they don't really think pure and applied are the same, but they don't care all that much. I think what I have just described is a fact, but I confess I can't help wondering why it's so.

New life

The deepest assertion about the relation between pure and applied mathematics that needs examination is that it is symbiotic, in the sense that neither can survive without the other. Not only, as is universally admitted, does the applied need the pure, but, in order to keep from becoming inbred, sterile, meaningless, and dead, the pure needs the revitalization and the contact with reality that only the applied can provide.

The first step in the proof of the symbiosis is historical: all of pure mathematics, it is said, comes from the real world, the way geometry, according to legend, comes from measuring the effect of the floods of the Nile. (If that's false, if geometry existed before it was needed, the symbiosis argument begins on a shaky foundation. If it's true, the argument tends to prove only that applied mathematics cannot get along without pure, as an anteater cannot get along without ants, but not necessarily the reverse.)

Insofar as all mathematics comes from the study of sizes (of *things*) and shapes (of *things*), it is true that all mathematics comes from the things of the real world. Whether renewed contact with physics or psychology or biology or economics was needed to give birth to some of the greatest parts of 20th century mathematics (such as Cantor's continuum problem, the Riemann hypothesis, and the Poincaré conjecture) is dubious.

The crux of the matter is, however, not historical but substantive. By way of a parable, consider chess. Mathematicians usually but sometimes grudgingly admit that chess is a part of mathematics. They do so grudgingly because they don't consider chess to be "good" mathematics; from the mathematical point of view it is "trivial". No matter: mathematics it is, and pure mathematics at that.

Applied mathematics cannot get along without pure, as an anteater cannot get along without ants, but not necessarily the reverse.

Chess has not been conceptually revitalized in many hundreds of years, but is vigorously alive nevertheless. There are millions of members of chess clubs, and every now and then the whole civilized world spends days watching Bobby Fischer play Boris Spassky. Chess fires the imagination of a large part of humanity; it shows them aesthetic lights and almost mystic insights.

Not only does chess (like many parts of mathematics) not need external, real-world revitalization, but, in fact, every now and then it spontaneously revitalizes itself. The most recent time that occurred was when retrograde chess analysis began to be studied seriously. (Sample problem: a chess position is given; you are to decide which side of the board White started from and whether Black had ever castled.) And here's a switch: not only was the real world not needed to revitalize chess, but, in fact, the life giving went the other way. Retrograde chess analysis challenged computer scientists with a new kind of problem, and it now constitutes a small but respectable and growing part of applied mathematics.

A revitalist might not be convinced by all this; he might point to the deplorable tendency of mathematics to become ultra-abstract, ultra-complicated, and involutedly ugly, and say that contact with the applications remedies that. The disease exists, that is well known, but, fortunately, so does nature's built-in cure. Several parts of mathematics have become cancerously overgrown in the course of the centuries; certain parts of elementary Euclidean geometry form a probably non-controversial example. When that happens, a wonderful remission always follows. Old mathematics never dies—what the Greeks bequeathed us 2500 years ago is still alive and true and interesting—but the outgrowths get simplified, their valuable core becomes integrated into the main body, and the nasty parts get sloughed off.

(Parenthetically: the revitalization argument could, in principle, be applied to painting, but so far as I know no one has applied it. Painting originates in the real world, it has been known to leave that world for realms of abstraction and complication that some find repulsive, but the art as a whole continues alive and well through all that.

A time argument is sometimes mentioned as a good feature of the contact of pure mathematics with applied. Example: if only pure mathematicians had paid closer attention to Maxwell, they would have discovered topological groups much sooner. Perhaps so—but just what did we lose by discovering them later? Would the world be better off if Rembrandt had been born a century earlier? What's the hurry?)

Whether contact with applications can prevent or cure the disease of elaboration and attenuation in mathematics is not really known; what is known is that many of the vigorous and definitely non-cancerous parts have no such contact (and probably, because of their level of abstraction, cannot have any). Current examples: analytic number theory and algebraic geometry.

When I say that mathematics doesn't *have* to be freshened by periodic contact with reality, of course I do not mean that it *must not* be: many of the beautiful concepts of pure mathematics were first noticed in the study of one or another part of nature. Perhaps they would not have been discovered without external stimulation, or perhaps they would—certainly many things were.

As far as the interaction between pure and applied mathematics is concerned, the truth seems to be that it exists, in both directions, but it is much stronger in one direction than in the other. For pure mathematics the applications are a great part of the origin of the subject and continue to be an occasional source of inspiration—they are, however, not indispensable. For applied mathematics, the pure concepts and deductions are a tool, an organizational scheme, and frequently a powerful hint to truths about the world—an indispensable part of the applied organism. It's the ant and the anteater again: arguably, possibly, the anteater is of some ecological value to the ant, but, certainly, indisputably, the ant is necessary for the anteater's continued existence and success.

What's next?

The most familiar parts of mathematics are algebra and geometry, but for the profession there is a third one, analysis, that plays an equally important role. Analysis starts from the concept of change. It's not enough just to study sizes and shapes; it is necessary also to study how sizes and shapes vary. The natural way to measure change is to examine the difference between the old and the new, and that word, "difference", leads in an etymologically straight line to the technical term "differential equation".

Most of the classical parts of applied mathematics are concerned with change—motion—and their single most usable tool is the theory and technique of differential equations.

Phenomena in the real world are likely to depend on several variables: the success of the stew depends on how long you cook it, how high the temperature is, how much wine you add, etc. To predict the outcome correctly, the variables must be kept apart: how does the outcome change when a part of the data is changed? That's why much of applied mathematics is inextricably intertwined with the theory of *partial* differential equations; for some people, in fact, the latter phrase is almost a synonym of applied mathematics.

Are great breakthroughs still being made and will they continue? Is a Shakespeare of mathematics (such as Archimedes or Gauss) likely to be alive and working now, or to be expected ever again? Algebra, analysis, and geometry—what's the mathematics of the future, and how will the relations between pure and applied develop?

I don't know the answers, nobody does, but the past and the present give some indications; based on them, and on the hope that springs eternal, I'll hazard a couple of quick guesses. The easiest question is about great breakthroughs: yes, they are still being made. Answers to questions raised many decades and sometimes centuries ago are being found almost every year. If Cantor, Riemann, and Poincaré came alive now, they would be excited and avid students, and they would learn much that they wanted to know.

In the foreseeable future discrete mathematics will be an increasingly useful tool in the attempt to understand the world, and . . . analysis will therefore play a proportionally smaller role.

Is there an Archimedes alive now? Probably not. Will there ever be another Gauss? I don't see why not; I hope so, and that's probably why I think so.

I should guess that in the foreseeable future (as in the present) discrete mathematics will be an increasingly useful tool in the attempt to understand the world, and that analysis will therefore play a proportionally smaller role. That is not to say that analysis in general and partial differential equations in particular have had their day and are declining in power; but, I am guessing, not only combinatorics but also relatively sophisticated number theory and geometry will displace some fraction of the many pages that analysis has been occupying in all books on applied mathematics.

Applied mathematics is bound to change, in part because the problems change and in part because the tools for their solution change. As we learn more and more about the world, and learn how to control some of it, we need to ask new questions, and as pure mathematics grows, sloughs off the excess, and becomes both deeper and simpler thereby, it offers applied mathematics new techniques to use. What will all that do to the relation between the ant and the anteater? My guess is, nothing much. Both kinds of curiosity, the pure and the practical, are bound to continue, and the Socrates of 2400 years from now will probably see the difference between them as clearly as did the one 2400 years ago.

So, after all that has been said, what's the conclusion? Perhaps it is in the single word "taste".

A portrait by Picasso is regarded as beautiful by some, and a police photograph of a wanted criminal can be useful, but the chances are that the Picasso is not a good likeness and the police photograph is not very inspiring to look at. Is it completely unfair to say that the portrait is a bad copy of nature and the photograph is bad art?

Much of applied mathematics has great value. If an intellectual technique teaches us something about how blood is pumped, how waves propagate, and how galaxies expand, then it gives us science, knowledge, in the meaning of the word that deserves the greatest respect. It is no insult to the depth and precision and social contribution of great drafters of legislative prose (with their rigidly traditional diction and style) to say that the laws they write are bad literature. In the same way it is no insult to the insight, technique, and scientific contribution of great applied mathematicians to say of their discoveries about blood, and waves, and galaxies that those discoveries are first-rate applied mathematics; but, usually, applied mathematics is bad mathematics just the same.

Solving Equations Is Not Solving Problems

Jerome Spanier

In 1955 I was a Ph.D. candidate at the University of Chicago completing a dissertation in Topology under Shiing-Shen Chern. Although I had been trained exclusively in pure mathematics, jobs in industry were beginning to become available to mathematicians and I was interested. Some of my teachers expressed disapproval of such notions, but Professor Chern did not. He told me he believed that working on physical problems was interesting and difficult, and he encouraged me to keep an open mind. I found that I was curious to learn more about the applications of the mathematics I had studied, so upon graduation I took an industrial rather than a teaching position.

The first six months were very difficult for me. I was anxious to begin the transition from pure to applied mathematician, but at a loss to know how. My supervision ranged from minimal to nonexistent (Ph.D. mathematicians were evidently expected to know what to do with themselves) so I was pretty much on my own. I read considerably and gained some satisfaction from what I was learning, but I seemed to be moving no closer to an encounter with "real problems."

I had nearly decided to give up and seek an academic appointment when I tried a different approach. My position was in a laboratory whose research was concerned with the then-blossoming field of atomic energy. I made a point of speaking to physicists, engineers, chemists—as many

Jerome Spanier is Professor of Mathematics at Claremont Graduate School. He received a B.A. in mathematics from the University of Minnesota in 1951 and a Ph.D., also in mathematics, from the University of Chicago in 1955. Subsequently, he worked for twelve years at Westinghouse's Bettis Atomic Power Laboratory, applying mathematics to problems of nuclear reactor theory. In 1963 he was awarded the Westinghouse Distinguished Service Award for "outstanding contributions to the improvement of nuclear reactor design capability." Following further industrial work at Rockwell's central research laboratory, where he became Group Leader of the Mathematics Group, Spanier joined the Claremont faculty where he has served as Co-director of the Mathematics Clinic since 1974.

different scientists as possible—about the nuclear reactor problems they were working on. Gradually I began to learn enough about reactor physics to appreciate some of these problems. My reading turned to physics and engineering more than to mathematics *per se*.

Finally, after more than a year, I felt ready (with the help of non-mathematician co-workers) to formulate and tackle some of the problems with which I had become somewhat familiar. This painful introduction to mathematical modeling was absolutely essential to my progress as an applied mathematician. My first job lasted nearly twelve years and the work I began then has continued as a research theme throughout my career.

If we want to teach students to apply mathematics effectively, we must do much more than just put techniques for solving various equations in their hands.

Of the 25 years which have passed since 1955, I spent the first sixteen working as an industrial applied mathematician and the last nine teaching applied mathematics. I have given considerable thought to doing and teaching applied mathematics. Although as a student I felt uncertain about the relation between mathematics and its applications, I am now absolutely convinced that the use of mathematics to solve real problems is an interesting and challenging process which leads to genuinely new mathematics every bit as valid as work in abstract mathematics. After all, ancient geometry was created, in part, to answer questions raised in measuring land areas, and information theory arose out of consideration of modern communication systems.

I am also convinced that if we want to teach students to apply mathematics effectively, we must do much more than just put techniques for solving various equations in their hands. In this essay I shall advocate that a liberal dose of applying mathematics ought to be an integral part of the education of every applied mathematician.

The nature of applied mathematics

In thinking about criteria for a proper education in applied mathematics, it seems appropriate to compare preparation for a more traditional mathematical career with what appears to be necessary training for work in the applications. Abstract mathematics proceeds from a system of axioms, or assumptions, to a collection of theorems (truths) and counterexamples (falsehoods) which can be established within such a system. Further work proceeds by extending and sharpening existing results, or by specializing to fit particular realizations. The teaching of techniques of proof plays an important role in training persons for work in abstract mathematics.

By contrast, work in applied mathematics begins with the statement of a problem which has arisen in a discipline outside of mathematics. The applied mathematician's first task is to create a mathematical image—or model—of the physical process which incorporates *realistic* assumptions and constraints. From this model, mathematical evidence is gathered which must then be compared with physical evidence typically gained through experiments or observation. This comparison determines the worth of the mathematical model. If the model is found wanting it must be altered and new mathematical evidence must be gathered. Quite clearly the original problem is of paramount importance to the effectiveness of the modeling process.

If one accepts these contrasting requirements, it seems evident that the teaching of applied mathematics calls for new approaches. Traditional courses in mathematics are technique-oriented; students are taught how to deal with a problem once it has been formulated mathematically. Thus, students are told how to solve algebraic equations, differential equations, integral equations—all kinds of equations—and how to formulate and prove theorems about such systems; but they are not taught where or how these equations arise. And yet the applied mathematician in industry or government must be prepared to help translate the engineer's or the economist's description of a problem into mathematical language *before* he can apply any of the ideas he has learned in traditional courses. The critical steps required to transform a concrete problem into mathematical symbols and equations, as well as the subtler techniques needed to evaluate the worth of the resulting mathematical model, are simply not treated in traditional courses.

Mathematical modeling

Partly to remedy this defect, it has become quite fashionable in the past decade to offer courses in mathematical modeling. A great variety of such courses has been introduced; the feature these seem to share is an emphasis on problem-orientation, rather than on technique. (See [1] for a description of one such course developed at Claremont Graduate School.) Students in these courses are exposed to "solved" real problems, ranging from those that are simply stated and accessible to undergraduates to complex problems arising in modern technology that require the most sophisticated mathematical treatment. Although such modeling courses may be excellent, they fail to give students the experience of tackling significant *unsolved* problems on their own except perhaps through projects undertaken as part of the course requirements. My contention is that when students are forced to develop their own approaches to unsolved problems they benefit much more than from exposure to the mathematics alone.

Although modeling courses can be excellent vehicles for acquainting students with some aspects of the work of an applied mathematician, they cannot really prepare students adequately for industrial problem-solving.

When students are forced to develop their own approaches to unsolved problems, they benefit much more than from exposure to the mathematics alone.

The reasons are, first, that industrial problem-solving almost always involves communication with non-mathematicians; second, it usually involves teamwork and shared responsibilities; and third, industrial work of a substantial nature requires a considerable depth of penetration into the discipline in which the problem arises. In many other respects as well, courses in modeling offer rather pale imitations of actual industrial problems.

A superior educational device in many respects is an applied mathematics practicum. This might range from an internship in which a student is introduced (normally individually) into a working environment, to a full-scale duplication of many aspects of the working environment in a classroom setting.

Internship vs. educational project activity

An internship might seem to provide an ideal way to introduce students to industrial work. However, I believe that although there may be some value in sending students to the working environment, there is much more to be gained in bringing the working environment into the classroom.

When a student accepts an internship, he may (and often is) given an assignment to assist an individual who acts as the student's supervisor. Rarely does the student become involved in substantial independent activity and almost never does he become part of a working team. He may be asked to do a fair amount of reading on his own and he will often be given a computer programming assignment. In a small number of instances he may be performing more nearly the work of a research assistant. While each of these activities is useful, the overall situation is less than ideal and the student is often bored and/or frustrated. Usually, no team activity is encountered, and (often) no real applied mathematics is learned. Total control of the educational experience is turned over to the sponsoring firm, and this does not usually serve the best interests of the student.

By contrast, accepting an industrial project activity as part of an ongoing curriculum allows the school to maintain educational control and often results in a superior experience. I shall describe one such practicum—The Mathematics Clinic—instituted at Harvey Mudd College (one of the six Claremont Colleges) in 1973 as a part of its training of applied mathematics students (see [2], [3], [4], and [5]). At the present time several other schools (for example, the University of California at Davis, Rensselaer Polytechnic Institute, Towson State University) have adopted similar plans.

What is the Mathematics Clinic?

The Clinic concept was created at Harvey Mudd College around 1964 to serve the needs of its engineering majors. A Mathematics Clinic was established at the same college in 1973-74; the following year responsibility to administer the Mathematics Clinic was undertaken jointly by Harvey Mudd and Claremont Graduate School. This unique course affords both undergraduate and graduate students an opportunity to work in teams on real problems originating in industry or government. Fixed fee research agreements are negotiated with the sponsoring firms before projects are undertaken. Funds from these provide some stipends for graduate students who function as "team leaders," and they also help support faculty project supervision as well as some of the administrative costs. The Clinic serves to carry the notion of mathematical modeling to its logical limit. Every attempt is made faithfully to recreate the industrial environment within the academic framework while exercising very careful control over the educational quality of the experience.

To date about 240 student-semesters have been invested in 61 semester-projects which have been tackled by student and faculty Clinic teams. These cover an impressive variety of topics, but methods developed to treat one problem may often be transferred to another.

Thus, for example, Clinic teams have accumulated more than 8 years of experience studying moving boundary/interface problems arising in connection with one or another nonlinear flow process. About half of this effort was expended helping Chevron scientists construct and analyze models for studying the abnormally high fluid pressures encountered underground in drilling for oil. Another two year project for Atomics International Division of Rockwell developed techniques for studying the sometimes disastrous swelling encountered in nuclear fuel elements, while two more years of effort were invested under the sponsorship of the Claremont water authority to aid in predicting groundwater nitrate concentrations (such nitrogen-bearing chemicals are known carcinogens and their control is of obvious importance). At the present time it appears likely that techniques similar to those used in these past projects will also be of value in helping the U.S. Forestry Service to understand the recovery of sedimentation rates in land which has been burned over by fire. Numerous other examples can be cited in which ideas developed in connection with one project have applicability in another project.

The Mathematics Clinic approach has many advantages, and reaction of the participants has been overwhelmingly positive so far. Students appear to enjoy the excitement and challenge of developing their own, often quite unique, solutions to genuine problems and communicating those solutions through talks and written reports to the sponsoring companies. The sponsors themselves receive the double reward of participating in a novel educational venture which they judge to be more meaningful than conventional ones, and reaping the benefits of an intensive mathematical analysis, professionally carried out, of a problem they face. Clinic participation also benefits both student and sponsor by providing contacts which can lead to jobs for students and reduced recruiting and on-the-job training costs for their employers.

A singularly important benefit of Clinic involvement is the confidence it gives students, confidence that comes from dealing with open-ended real problems.

Two National Science Foundation grants have helped the Mathematics Clinic grow to a stable, self-sustaining size. One of these grants has initiated a post-doctoral training program in which mathematicians, through a combination of traditional course work and Clinic consulting, are retrained for a year in applications. Other applied mathematicians have visited Claremont for periods of six months to two years to learn first-hand of the Clinic program and to gain experience working in it. A total of 15 such persons has been attracted to Claremont in the past three years. A number of these visitors are already introducing elements of this novel program at other institutions.

A singularly important benefit of Clinic involvement is the confidence it gives to students, confidence that comes from dealing with open-ended real problems. By exposing students to at least one paradigm of applied research in their own project and to others by having to listen to oral reports on additional projects, and by sharpening their communication skills (which employers find to be invaluable), the Clinic provides concrete preparation for the students' future careers as problem-solvers.

Nor have the benefits derived from Mathematics Clinic work been bestowed only upon students interested in industrial careers. Most of Claremont Graduate School's Ph.D. candidates in the past six years have involved themselves in Clinic project activity at some time. This work has helped them to broaden their educational base and to acquire problem-formulation and problem-solving skills which have served them well in their chosen careers, whether in industry or in academia. Faculty, too, have grown through their Clinic experiences. Exposure to industrial and governmental problems has given them the raw material to become more effective teachers by injecting elements of their experiences into more standard courses. Furthermore, in some cases, opportunities for summer consulting positions have arisen which are both financially rewarding and helpful in identifying new areas for future project activity.

Requirements for training applied mathematicians

In my analysis of requirements for training students for work in the applications of mathematics, I do *not* advocate wholesale replacement of fundamental mathematics courses by modeling and practicum work. Frankly, my own view is that the successful applied mathematician must know as much mathematics as possible and must also acquire some experience, of the sort I have described, in dealing with open-ended real problems. The traditional and the new educational elements need to be related skillfully for maximum benefit.

A graduate-level modeling course [1] and a unique practicum in which students grapple with real problems lie at the heart of Claremont's new program in applied mathematics. The program has been effective in attracting students and in placing them after graduation. Experience gained in developing and testing this novel program suggests that the following constitute very important requirements in training applied mathematicians and prove to be valuable even more generally:

(1) a focus on problem-solving;
(2) experience in communications, both oral and written;
(3) familiarity with cognate disciplines;
(4) exposure to at least one paradigm of applied mathematics research;
(5) confidence acquired in open-ended problem solving.

While many of the specifics of the program I have described may not be easily duplicated elsewhere, some features of it deserve strong consideration by any institution that would train students to use mathematics effectively.

References

[1] Jerome Spanier. "Thoughts About the Essentials of Mathematical Modeling." *Int. J. Math. Modeling* 1 (1980) 99-108.
[2] ———. "The Claremont Mathematics Clinic." *SIAM Rev.* 19 (1977) 536-549.
[3] ———. "The Education of a Mathematical Modeler." *Proc. Second Int. Conf. on Math. Modeling*, St. Louis, July 11-13, 1979.
[4] Melvin Henriksen. "Applying Mathematics Without a License." *Amer. Math. Monthly* 84 (1977) 648-650.
[5] Robert L. Borrelli and Jerome Spanier. "The Mathematics Clinic: A Review of its First Seven Years," *UMAP J.* (forthcoming).

The Unexpected Art of Mathematics

Jerry P. King

There are moments so rare as to bring with them a different kind of time. An event occurs so charged with emotion and intensity that one's biological clock stops, the background fades away, and the thing itself is seen frozen and close-up as through a zoom lens. You can hear the turn of the key as the scene locks itself deep inside your brain.

And, instantly, there is an awareness of a kind of reverse *déjà vu*. You know that, for as long as you live, this moment will recur and recur and each time you will pull it from your memory it will gleam like a gold coin from a velvet case.

Many of these moments are associated with public events. You remember exactly where you were when you learned of the assassination of John Kennedy or the death of Franklin Roosevelt. And, even though you were six years old, you will forever remember the bombing of Pearl Harbor. On that Sunday afternoon the grownups gathered in the parlor and talked quietly and somberly. Something had frightened them. Badly. And, when the grownups were frightened, so were you.

But there are private moments of the same kind. Once, thirty years ago, a pass was thrown to me in a high school football game. I can blink my eyes and bring the scene back. I'm running in the endzone, as wide open as a beach umbrella. The ball comes to me on a soft lazy spiral. It is timed perfectly with my motion. No thrown football was ever more catchable. It smacks gently into my palms. Spectators leap to their feet and cheer, but I do not hear them. I cradle the ball and tuck it away. Then, inexplicably, the ball pops loose and I am on my knees, sliding on the grass, scrambling after

Jerry P. King is Professor of Mathematics and Dean of the Graduate School at Lehigh University. He holds a bachelor's degree in electrical engineering and in 1962 he received a Ph.D. in mathematics from the University of Kentucky. His research has been in complex analysis and summability theory. King was a member of the Board of Governors of the Mathematical Association of America during 1977-80. In 1976 he won Lehigh University's Stabler Award for Excellence in Teaching.

it. I stretch desperately toward the tumbling football, as if, by touching it, I can make the thing right, save the touchdown, win the game. But the ball rolls away.

Hardly a week goes by but what I still reach for that football. I didn't touch it then and I will touch it never. But I go on reaching.

What all these events have in common is their unexpectedness. You did not *expect* to hear of the assassination or of the bombing. You did not *expect* to drop the ball in the endzone. Each event is as completely unexpected as a ghost at the top of the stairs. Or as pure mathematics in an ordinary classroom.

Pure mathematics

One can do worse than quote Paul Halmos on mathematics. He said: "It saddens me that educated people don't even know that my subject exists."

Halmos practices the profession of pure mathematics, which is mathematics for mathematics, as opposed to applied mathematics, which is mathematics for something else. He is, of course, entirely correct when he says that most educated people do not know that pure mathematics exists. But there was a time, not long ago, when things were worse. There was a time when even people educated in *mathematics* did not know that pure mathematics existed. And there is every indication that those times will return. Slouching toward us is a new curriculum whose spirit is as contrary to pure mathematics as a nightmare to a dream. When it comes, Halmos' sadness will change to deep lamentation.

Pure mathematics is mathematics for mathematics, as opposed to applied mathematics which is mathematics for something else.

Most of the mathematicians of my generation will admit they came to their profession by accident. Few of us set out to be mathematicians. Indeed, because all of our secondary school and beginning college courses in mathematics had been taught from the point of view of applications, none of us knew that such a profession was possible. Because we had shown some competence in high school mathematics we had been advised to become engineers or scientists. Our teachers, having themselves been taught mathematics only as a tool for applications, had no notion that someone good at mathematics might consider becoming a mathematician.

So we began, in college, to study engineering. We took calculus, differential equations, and linear analysis, all of these being courses emphasizing specific mathematical techniques useful in engineering applications. Then, just by chance, we took a course outside our curriculum. We enrolled in a post calculus course in analysis. And, unexpectedly, we encountered pure mathematics. We were *struck* by the subject, like Saul on the road to Damascus.

Nothing in our backgrounds had prepared us for the aesthetics of mathematics. We saw, for the first time, a professor who treated mathematics with reverence, who wrote symbols on the blackboard with great care as if they mattered as much as the information they contained. We heard a mathematical result described as "elegant." And we saw that it was.

It was a moment of great discovery. We felt as if we had lived our lives in the hold of some great ship and now we were brought on deck into the fresh air. And we saw the unexpected sea. We felt, with W.B. Yeats:

All changed, changed utterly:
A terrible beauty is born.

The moment was worth the wait but we would be forever aware we had come to it entirely by chance. And, along the way, there had been many drop-outs.

Trivialities

In secondary schools, we had taken a succession of courses which consisted mainly of the repetitive manipulation of mathematical symbols. And always each dull course had been justified on the grounds that, once it was mastered, we would have at hand a kind of machine for transportation. This machine, they told us, would transport us from the classroom to some vague place called the "real world." And the word they used most frequently to indicate the link between this place, wherever it was, and mathematics was "applications."

As an application of algebra we had computed the age of a farmer who is twice as old as his son will be in six years if the farmer is now three times as old as the son. Using geometry, we had measured the size of farm plots along the Nile River. Trigonometry had been "applied" to the determination of the heights of an unending succession of trees, once certain angles were measured.

But we didn't care much about the age of fictitious farmers or the size of farm plots or the heights of trees. Even those of us with scientific interests were not persuaded there was any connection between such calculation and the difficult notions we believed lay ahead in physics or chemistry. We were naively certain that, in order to get at the central core of scientific truth, one needed a tool at least as complicated as a jackknife. And all we had seen of mathematics was its application to a dull sequence of trivialities.

For the others, the non-scientists, mathematics was something to be endured. What interested them about farms and rivers and trees had nothing to do with calculation. They cared for pasture springs, the sound of wind through pines, the slant of sunlight on bright water. Mathematics was to them as far removed from aesthetics as is a dusty toolbox from a polished violin.

These humanists waited. They patiently suffered the minimal amount of mathematics instruction and took silent vows that, when it was over, they would never again allow the subject brought into their presence.

The rest of us continued haltingly through the curriculum. We sidled up to each mathematics course the way a kid with skinned knees limps toward his two day old bicycle. Mathematics might well be a vehicle as our teachers said, but it had, so far, taken us nowhere we wanted to go. So far, mathematics had only thrown us on our knees.

College calculus came next and with it the defeat of all expectation. Through high school, calculus had been dangled before us like some golden ring. Our teachers had talked of it as if it were the capstone of all mathematical knowledge. There was never the slightest indication that analysis beyond calculus existed. "When you get to calculus," they said, "you will see what mathematics is really good for."

What we saw was yet another course in manipulation taught from the point of view of physical applications. Derivatives were presented as velocities and integrals as areas. As far as we knew, the difference between a Riemann integral and an anti-derivative was that one of them was evaluated between two limits. And for the purposes of the course no deeper knowledge was needed.

We waited a full year before we, unexpectedly, discovered the existence of pure mathematics and learned that elementary calculus had been given to us exactly the wrong way around. The truth is that there is a mathematical notion of the derivative of a function and it sometimes can be interpreted as a velocity. But derivatives also have interpretations as, for example, interest rates or probability densities or population growths. So what one should study are derivatives and then the applications appear as special cases.

Suddenly we understand that mathematics possesses an aesthetic value as clearly defined as that of music or poetry.

One full year passed after elementary calculus before we learned the true relationship between Riemann integrals and antiderivatives, and we discovered that the connection between these different notions lies at the very heart of the subject. We saw that the argument establishing this link was a thing of great beauty. And suddenly we understood that mathematics possesses an aesthetic value as clearly defined as that of music or poetry.

Calculus

Fashions change in mathematics teaching just as they change in the teaching of anything else. Those of us who have watched the evolution of calculus books over the last thirty-years have seen a classic example of this change.

In the beginning, calculus texts were relatively thin and were separated into two distinct parts, one called differential calculus and the other called integral calculus. In the 1950's these texts were replaced by thick volumes whose titles were all some variation of the phrase "calculus with analytic geometry." These newer books presented the topics of both differential calculus and integral calculus in a more or less homogenized manner and they incorporated basic material on analytic geometry that was formerly taught in precalculus courses. These texts were followed by other books called "calculus with linear algebra" and then by a short, quick burst of books called "calculus with probability."

And, at each stage of this progression, authors and publishers eagerly pointed out the enormous pedagogical advantage their present book had over all its predecessors. These exaggerations of value were made regularly, one after another, with the same disregard for logic that once led twenty-nine year old radicals to shout the slogan "don't trust anyone over thirty."

But the changes, except for the first one, were insignificant. The "linear algebra" and "probability" texts were as similar to George Thomas's 1951 *Calculus with Analytic Geometry* as a sea lion is to a seal. The important change was the initial one, the abrupt leap from the old differential calculus-integral calculus textbooks to the book of Thomas. And, while most of the mathematicians of my generation studied elementary calculus from texts of the earlier type, we began our teaching with Thomas.

Two factors distinguished Thomas's book from its predecessors. On the one hand, the book contained much more material than did the older texts. There was the additional material on analytic geometry to be sure but, more importantly, it contained topics on vectors, infinite series, and differential equations that were traditionally taught in more advanced courses.

Also, Thomas's book differed from many previous calculus texts in that its tone was much closer to the tone of pure mathematics. Most of the traditional applications were included, but the combination of advanced topics and increased rigor gave the students an inkling that there might be more to mathematics than its use in computing moments of inertia.

Thomas's book and the wave of imitators which followed gradually took hold of the elementary college curriculum. As they did, there began a movement of the standard college calculus course in the direction of pure mathematics. The move was minuscule and barely perceptible. Left alone, it would have crawled to a stop like a snail in the hot sun. But help came.

On October 4, 1957 the Soviet Union successfully launched a 183 pound satellite called Sputnik I. A startled nation reacted. Money for basic research flowed from federal coffers. The golden age of pure mathematics began.

For the next twenty years, college students in the United States would learn calculus from texts whose sap and root was that of Thomas. For a time, American colleges and universities produced students who, while learning calculus, gained some vague notion of the existence of Paul Halmos's subject.

The age of applications

Times change. We are now deep into the age of applications. A new generation of calculus texts has appeared. These new books have titles such as "applied calculus" or "goal oriented calculus" or "calculus with applications." They emphasize the applicability of calculus to the real world, often through problems associated with everyday situations, rather than through the more esoteric physics-mechanics exercises of earlier texts.

Times change. We are now deep into the age of applications.

This turn toward applications is dramatic and pervasive. The entire mathematics curriculum has been changed. And the shift has been noticed by the experts. They have looked upon it and found it good. But not good enough.

The prestigious National Research Council, through its Committee on Applied Mathematics Training has put forward a report called "The Role of Applications in the Undergraduate Mathematics Curriculum." The thrust of their recommendations is that both students and society will benefit from a mathematics curriculum which emphasizes applications.

The Committee on the Undergraduate Program in Mathematics, a group associated with the Mathematical Association of America, will soon produce recommendations on a "General Mathematical Sciences Program." The expectation is that C.U.P.M. will recommend a radical revision of the traditional undergraduate mathematics program. They will recommend that the traditional mathematics major be replaced by a major in mathematical sciences which has as its central theme applications of mathematics to the real world.

These are influential organizations. Their recommendations have great weight. There is no turning back.

> Things fall apart; the centre cannot hold;
> Mere anarchy is loosed upon the world.

The age of applications must run its course.

Two recommendations

Neither the National Research Council nor the Committee on the Under-graduate Program in Mathematics has asked me to advise them as to their recommendations on applications of mathematics. I assume this to be a mere oversight, since it is well-known that giving advice is one of the things I do best. So I will advise them now.

My advice comes in two parts and it is free. They can decide what it is worth.

1. *The efficacy of applications as motivation should not be exaggerated.*

It is fashionable to argue that students will study mathematics with more enthusiasm when the subject is presented from the point of view of applications than when it is not. My experience, both as a mathematics student and as a teacher, is that this is not unequivocally true.

Unquestionably, a *particular* student can be motivated to study a *particular* piece of mathematics by showing him how the mathematics is applicable to a problem in which he or she is interested. A student who wants to learn classical mechanics, for example, can be motivated to study ordinary differential equations through exposure to the contrasting mathematical models of a falling raindrop and a falling ping-pong ball. The student might be asked to invent these models for himself, using only Newton's Second Law and his intuition as to what the resistive forces might be. Moreover, the student might be asked to use his intuition to make rough guesses as to the form of the solutions of the differential equations in the various cases.

There can be no doubt that this process is instructive and motivational for a student who *cares* about falling bodies. But it is also a highly individualistic and inefficient way of introducing differential equations. And, unfortunately, not every student is interested in falling bodies. A student who wants to study economics, for example, might be as bored with this application as he was with the tiresome surveying exercises of trigonometry and plane geometry.

Of course, students can be separated by fields of interest and then taught mathematics. This is, in fact, a common practice in the teaching of calculus. But once this is done, little motivation is necessary. Nobody needs to motivate a classroom full of electrical engineers to study calculus. They are already highly motivated. The motivation is the next course.

But there is a large group of students who cannot be reached by applications of any kind. These are the ones who will eventually wind up on the left bank of the river separating C.P. Snow's infamous "two cultures." In the 1959 Rede Lecture, Snow called these people the "literary intellectuals." They are now generally referred to as "humanists."

No application of mathematics whatever appeals to the humanists. They are interested neither in the bending of beams nor in the oscillation of pendulums. They care deeply for the Heifetz version of the Bach *Chaconne* but are turned off by mathematical analysis of vibrating strings. You can interest them in health problems associated with nuclear power but you cannot motivate them to study equations of radioactive decay.

That is, you cannot motivate humanists to learn mathematics on the grounds of its practicality, its utility, or its applicability. But it is possible to engage them with mathematics. What you do is emphasize the aesthetics of mathematics. You show them mathematics as a classical and intellectual art. And you do it early, before the barrage of applications drives them away.

2. *The efficacy of aesthetics as motivation should not be underestimated.*

Let us hear two mathematicians and a physicist on mathematics and aesthetics. The mathematicians are Henri Poincaré and G.H. Hardy, the physicist is Werner Heisenberg. Poincaré's remark was made in his celebrated lecture before the Société de Psychologie in Paris. Hardy's statement is from his essay *A Mathematician's Apology*. Werner Heisenberg's remark was made in conversation with Albert Einstein.

Finally, we listen to John Keats who is talking about a Grecian urn.

Poincaré: "To create consists precisely in not making useless combinations. ... The useful combinations are precisely the most beautiful."

Hardy: "Beauty is the first test: there is no permanent place in the world for ugly mathematics."

Heisenberg: "If nature leads us to mathematical forms of great simplicity and beauty ... we cannot help thinking that they are true, that they reveal a genuine feature of nature."

Keats:

> Beauty is truth,
> truth beauty—that is all
> Ye know on earth,
> and all ye need to know.

I do not expect to convert anyone with recommendation number two. The pure mathematicians are already believers. Indeed, it is exactly the aesthetics of mathematics that attracted them to the subject in the first place. The applied mathematicians who do not believe it are not likely to change their minds at this late date. And for the rest, those multitudes for whom mathematics is at best a painful memory, the very notion that it can have *any* aesthetic value is as fantastic as the idea that pigs have wings.

The most I can accomplish with the second recommendation is a weak compromise. The applications enthusiasts hold all the cards. They have behind them the power and the influence of the national organizations and commissions. They are reshaping the mathematics curriculum in their own image. There is no chance I can change their minds.

But I ask a favor. Let one course, just one, remain pure. And let it be beginning calculus.

This will cause no shortage of applications. In fact, in the new curriculum, each course beyond calculus will be rich with them. The students will study computing, applied combinatorics, applied statistics, and operations research. All of these courses will be as smeared with applications as the night sky with stars.

They will see pure mathematics. And they will never care for anything half as much.

But we owe the students one course where they see mathematics taught as mathematics by a pure mathematician. And it must be an elementary course. Otherwise, the humanists will have already dropped out and an opportunity will have been missed. An opportunity to bring them to mathematics through aesthetics and thereby causing a slight narrowing of the gap between the two cultures.

And we owe Paul Halmos a chance to see that some mathematics students know that his subject exists.

I am saddened by the new curriculum but I am not in despair. I will go on pretty much as before. Most of my own post-calculus teaching has been in courses designed for engineers. They contain many applications and will not change much.

As for my other courses—upper division courses in complex analysis, real analysis, mathematical probability—I am resigned to seeing them dwindle in enrollment and then, one-by-one like blown candles, disappear. They cannot survive indefinitely alongside the applied mathematics curriculum.

But they will last long enough for me. I can always find just one more pure mathematics course to teach, the way an aging gunfighter, with times changing, and law-and-order settling in, always found one more town to tame.

Just one more course and I'm done. Make it classical complex variables. Let me do it once more.

And one day when the wind is right I'll do the Cauchy Integral Formula for the last time and I'll do it truly. I will write it carefully and the students will see the curve and the thing inside and the lazy integral that makes the function value appear as suddenly as my palm when I open my hand.

They will see pure mathematics. And they will never care for anything half as much.

Redefining the Mathematics Major

Alan Tucker

Mathematical methods and thinking permeate virtually all aspects of business, government, and academia today to an extent few could have imagined a generation ago. This growth is manifested in the omnipresent role of computers in today's world and in the use of mathematical models and statistical analysis to plan everything from medical treatment to political speeches. Mathematical terms, such as "parameter," "hypothesis," and "unknown variables," have become part of the business jargon. It used to be that most college students took mathematics courses solely for the intellectual discipline of studying mathematics. (Many schools required every student to take either two years of mathematics or two years of a foreign language.) Today, courses in the mathematical sciences are an integral part of most college majors. Business students, for example, are frequently required to study statistics, linear programming, and calculus because this mathematics and its associated modes of reasoning are widely used in management.

Yet while overall college mathematics enrollments have been steadily increasing, the number of mathematics majors has been declining for the past decade. This paradox can be ascribed to many factors, but the fundamental reason, in my view, is that the types of mathematical methods and thinking that society needs have been neglected in the traditional

Alan Tucker is Chairman of the Department of Applied Mathematics and Statistics at the State University of New York at Stony Brook. He received a B.A. in applied mathematics from Harvard University in 1965 and a Ph.D. in mathematics from Stanford University in 1969. Since joining the Stony Brook faculty, he has developed an undergraduate program in applied mathematics and statistics that has served as a model for many other schools. He is currently chairman of the Mathematical Association of America Panel on a General Mathematical Sciences Program; this panel is revising the Association's current model mathematics major to become a major in mathematical sciences. Tucker's research has been in pure and applied combinatorial mathematics, including combinatorial algorithms and applications to optimizing municipal services.

mathematics major. Many efforts are now underway to restructure the mathematics major, to integrate modern applied mathematics with associated models and problem-solving into the standard analysis-and-algebra based curriculum emphasizing abstraction and proofs.

How past successes led to today's problems

Until the 1950's, the main career for students majoring in mathematics was teaching mathematics to others. Some mathematics graduates also found employment as actuaries in insurance companies. (Actuaries determine from life tables the proper premiums on insurance policies and perform related financial computations.) Mathematics departments were primarily service departments, teaching necessary skills to science and engineering students and teaching mathematics to most students solely for its liberal-arts role as a valuable intellectual training for the mind. The average student majoring in mathematics at a better college in the 1930's took courses in higher (college) algebra, trigonometry, and analytic geometry in the freshman year followed by a sophomore year of calculus. More calculus followed in the junior year (infinite series, multiple integration, ordinary differential equations), and the mathematics major was typically filled out with five or six electives in theory of equations, mathematics of finance, projective geometry, differential equations, logic, history of mathematics, number theory, probability and statistics, complex variables, and what is today called advanced calculus (e.g., continuity theorems, implicit function theorem, elements of Riemann-Stieltjes integration). Most mathematics majors also took a substantial amount of physics and chemistry. During this same period the training of secondary school mathematics teachers at state teachers colleges rarely included more than a year of calculus. Twenty years later in the mid-1950's, the situation had changed little, although the top universities did now offer modern algebra and abstract analysis. When the Mathematical Association of America (MAA) moved into the field of curriculum development in 1954, the first goal was defining a one-year sequence called Universal Mathematics in which the first semester was basically pre-calculus continuous mathematics.

The orbiting of Sputnik in 1957 launched mathematics education into the national limelight. Studying mathematics and science was valued by parents, employers, and society as a whole. The increased enrollments in secondary and college mathematics courses called for an even greater increase in the number of mathematics majors to fill the demand for secondary and college mathematics teachers, as well as for industrial mathematicians. The industrial mathematicians were used in large measure to do scientific programming of the new electronic digital computers. In 1956, there were 5000 Bachelors degrees awarded in mathematics, 1000

Masters degrees, and 250 Ph.D.'s. By 1970, the corresponding numbers were 27,000 Bachelors, 5,500 Masters, and 1250 Ph.D.'s.

There were three themes in mathematics curriculum seen by MAA curriculum planners. The first theme was a "traditional" set of courses emphasizing differential equations and mathematics' close historical bonds to physics. This theme was suited to students training for industrial careers and was also reasonable for secondary school mathematics teachers (who would teach mathematics to students interested in scientific careers). The second theme was a more modern "pure"-ish mathematics emphasizing newly developed theoretical foundations and generalizations of calculus and geometry. This theme was designed especially for students planning graduate study in mathematics and eventual careers in college teaching. Because some of the concepts of modern mathematics were being introduced into secondary school mathematics, this second theme also had value for secondary school teachers. Finally there was a theme in the formative stages involving the mathematics of "organized complexity" (to use a term coined by Warren Weaver). While calculus and kindred physical laws give a simple, global description of most physical phenomena, mathematical subjects such as statistics, operations research, and computer science involve complex structures of information that cannot be simply described. Undergraduate courses in this area were limited in the mid-1950's. The MAA curriculum group developing the Universal Mathematics sequence in 1954 anticipated the future importance of this mathematics by devoting the second semester in the sequence to a new "discrete mathematics" course, covering set theory and applications to probability and statistics.

Of these three themes for a mathematics major, the second one emphasizing pure mathematics became dominant in the 1960's. While the emphasis on graduate school preparation in the mathematics major may now seem to have been inappropriate for the "average" student, the fact is that most mathematics students then were happy with this program (and of course, the faculty liked it since it reflected their personal interests). The students came out of high school better prepared than today's students, ready to work hard to master this highly valued "language of science." Moreover the brightest students were going into mathematics and physical sciences. During the 1960's, the development of finite mathematics courses was largely ignored; such courses were no longer even recommended for mathematics majors.

The New Math was introduced in the late 1950's to bring fundamental concepts of pure mathematics and finite mathematics into the secondary curriculum. For example, sets and inequalities were introduced in large measure for their use in probability. However, in the 1960's, the New Math, like the college mathematics major, came to emphasize only the values of pure mathematics.

By 1970, the Vietnam War had started an anti-science/math trend among college students and the space program was in sharp decline. Many students were also anti-business. These attitudes undermined potential interest both in physical science mathematics and in finite mathematics, because of its applied, industrial uses. Nevertheless, compared to the severe enrollment declines registered in the physical sciences (including engineering) around 1970, mathematics was still doing well. The applicationless, mathematics-for-its-own-sake nature of a pure mathematics major and the availablity of teaching positions may have been the reason for this continued strength.

In the late 1970's, however, the mathematics major literally fell apart. Pure mathematics graduates were unwelcome in industry. There were few jobs in teaching, especially for Ph.D.'s. Industry wanted problem-solvers, not theorists, and fiscal constraints restricted the hiring of new teachers. Students were graduating from high school knowing less mathematics and with less disciplined study habits. As a consequence, the average students were getting much less out of the 1960's mathematics major because it was over most of their heads. Concern for industrial employment was sending many students with good mathematics skills to engineering and computer science, majors based on physical science mathematics and finite mathematics, respectively.

Finally it is worth noting that most mathematicians were assuming as late as 1970 that the rapid growth of the early 1960's in mathematics Bachelor and Ph.D. degrees would continue through the 1970's, because industry would want more mathematicians (pure or applied) and students at all levels would study more mathematics requiring more mathematics teachers. A 1970 projection of over 50,000 mathematics graduates in 1975 was a poor estimate of the 18,000 Bachelor degrees actually produced. Such optimistic predictions had led mathematics departments to neglect the teaching of applied subjects such as statistics and computer science. By 1975, the majority of statistics enrollments were in courses taught in social or biological sciences departments, and separate computer science departments were sprouting everywhere. A pure mathematics orientation in college mathematics teaching was now a liability, hurting both the traditional service function of mathematics departments as well as the training of mathematics majors.

Current challenges for the mathematics major

The difficulties in the mathematics major in the early 1970's have evolved in the past few years into a more complicated mixture of new problems and new opportunities. The unprecedented growth of computer science as a major new college subject is a prime example of this minus-and-plus situation. The mathematically oriented students who previously looked to secondary school teaching as a natural career now are thinking about computing careers. The number of computer science majors now exceeds the number of mathematics majors at most schools offering majors in both subjects. Enrollments in most upper-level pure mathematics courses declined dramatically in the 1970's as mathematics students turned to applied and computer-related courses. Yet while the number of mathematics majors is decreasing, the demand for broadly trained mathematics graduates is increasing in government and industry. They are needed to solve previously intractable problems now made accessible with modern digital computers. The mathematical problems inherent in projects such as optimizing the use of scarce resources or making operations in the public and private sector more efficient guarantee a strong future demand for mathematicians. These problems require people trained in disciplined logical reasoning and versed in the basic techniques and models of applied mathematics. If mathematics departments do not train these quantitative problem-solvers, then departments in engineering and management science will.

The shortage of secondary mathematics teachers has returned and now is worse than ever before. This shortage is being compounded by high-paying computing jobs that are attracting current mathematics teachers out of the classroom. Although the training of future teachers should include coursework in computing and applications, such coursework heightens the probability that these students will switch to careers in computing.

The mathematical problems inherent in projects such as optimizing the use of scarce resources or making operations in the public and private sector more efficient guarantee a strong future demand for mathematicians.

On another front, pre-calculus enrollments are soaring, while the mathematical skills of incoming freshman have been declining for several years (a problem that concerned the MAA's Universal Mathematics planners in 1954). The mathematics curriculum may soon need to allow for majors who do not begin calculus until their sophomore year, as was common a generation ago.

At universities, the decline in graduate enrollments has frequently overshadowed the decline in undergradute majors. Faced with heavy pre-calculus workloads, shrinking graduate programs, and competition from other mathematical sciences departments, university mathematics departments appear less able to broaden and redefine the mathematics major than do most liberal arts college mathematics departments. Many university mathematicians prefer to retain their current pure mathematics major for a small number of talented students rather than to redirect it to meet the diverse needs of today's average student.

Several encouraging developments are emerging, however. A natural evolution in the mathematics major is taking place at many schools. In some schools students and faculty have developed an informal "contract" for a major that includes traditional core courses in algebra and analysis along with electives weighted in computing and applied mathematics. Another favorable trend is the emphasis on symbolic, as opposed to numerical, computation in the current computer science curriculum. The Association for Computing Machinery (ACM) Curriculum 78 report transfers primary responsibility for teaching numerical analysis, discrete structures, and computational modeling (such as simulation) to mathematics departments. The ACM curriculum implicitly encourages students who are interested in scientific computation or in mathematical problem-solving that uses computers to be mathematics majors.

Some successful mathematics programs

In recent years I have been seeking out successful mathematics programs in connection with an MAA curriculum project. "Successful" means attracting a large number of students into a program that develops traditional mathematical reasoning and also offers a spectrum of (well taught) courses in pure and applied mathematics. The set of colleges mentioned here is only a sampling of the successful programs that have come to my attention. While mathematics majors now represent slightly above 1% of all college graduates, the successful programs typically produce 5% to 8% of their institution's graduates. Thus, a significant fraction of mathematics graduates nationally are now coming from this comparatively small number of successful mathematics programs.

Saint Olaf College, a 3000-student liberal arts college in Northfield, Minnesota, has a contract mathematics major. Each mathematics student presents a proposed contract to the Mathematics Department. The contract consists of at least 9 courses; college regulations limit the maximum number of courses that can be taken in one department to 14. The department normally will not accept a contract without at least one upper-level applied and one upper-level pure mathematics course, a computing course or evidence of computing skills, and some sort of independent study (research program, problem-solving pro–seminar, colloquium participation, or work-study). Frequently a student and an advisor will negotiate a proposed contract. For example, a faculty member will try to persuade a student interested in scientific computing and statistics that some real analysis and upper-level linear algebra should be included in the contract; this material is needed for graduate study in applied areas, and in any case a liberal arts education entails a more broadly based mathematics major. Conversely, a student proposing a pure mathematics contract would be confronted with arguments about not being able to appreciate theory without knowledge of its uses. In the end, the student and the faculty member understand and respect each other's point of view. The Mathematics Department offers minors in computing and statistics, but the attractiveness of a contract major in mathematics leads most students interested in these areas to eventually become mathematics majors.

A significant fraction of mathematics graduates nationally are now coming from a comparatively small number of successful mathematics programs.

Lebanon Valley College a 1100-student liberal arts college in Annville, Pennsylvania, has only 5 mathematics faculty but offers tracks within the mathematics major for mathematics graduate study, secondary teaching, and operations research, along with a separate major (run by the Mathematics Department) in actuarial science; in addition, a computer science major is about to start. Most of the upper-level coursework in the mathematics graduate preparation track involves problem seminars and formal and informal topics courses (because of the limited demand in this area). All mathematics majors must take a rigorous 22 semester-hour core of calculus, differential equations, linear algebra, and foundations. Despite a 40% attrition rate in mathematics and a 70% attrition rate in actuarial sciences, these two programs constitute 8% of Lebanon Valley graduates (mathematics graduates outnumber actuarial science 3 to 1). The Lebanon Valley mathematics faculty has recreated the pro-mathematics student attitudes of the 1960's by publicizing the attractive well-paid careers in the mathematical sciences to undergraduates and high schools in the region. Many students are initially attracted by the major in actuarial science (an historically established profession) and then move into other areas of applied and pure mathematics. Once the faculty have the students' "attention," they work the students hard. Because there are known rewards waiting for those who do well in mathematics, as well as a personal sense of intellectual achievement, the students respond positively to the demands of the faculty.

Nearby Gettysburg College has a special vitality to its mathematics program that comes from an interdisciplinary emphasis. The department has held joint departmental faculty meetings with each natural and social science department at Gettysburg to discuss common curriculum and research interests. Several interdisciplinary team-taught courses have been developed, such as a course on symmetry taught jointly by a mathematician and a chemist. An interdepartmental group organized two recent summer workshops in statistics which drew faculty from eight departments. Mathematics faculty have audited a variety of basic and advanced courses in related sciences to learn to talk the language of mathematics users. Mathematics faculty bring this interdisciplinary point of view into every course they teach, giving interesting applications and showing, say, how a physicist would approach a certain problem. Needless to say, a majority of mathematics majors at Gettysburg are double majors.

Small Goucher College near Baltimore, Maryland has achieved excellent unification in its mathematics program. Goucher has maintained a healthy interest in pure mathematics by integrating applications and computing into most mathematics courses. The unity of pure and applied mathematics is personified to the students by faculty members who teach both pure and applied courses with equal enthusiasm.

Usually a separate computer science department with its own major spells disaster for the mathematics major at a school. But remotely situated Potsdam State College (in the economically depressed northeast corner of New York state) has, to my knowledge, the largest percentage of mathematics graduates of any public institution in the country—close to 10%—despite competition from a popular computer science major. The Potsdam success is based on the same approach used at Lebanon Valley. The mathematics department publicizes the rewarding industrial careers available to a broadly trained mathematician, and once motivated the students are given a solid dose of pure and applied mathematics. There is a very active mathematics club to sustain and further publicize interest in mathematics. Now Potsdam attracts mathematics students from all over New York state because students have heard that Potsdam will prepare them to get good industrial jobs.

Directions for a healthy future

The success of the flourishing mathematics programs mentioned in the previous section points first of all to the importance of the right attitudes among the mathematics faculty. The faculty in these programs are deeply involved in undergraduate mathematics education. They have realized the fundamental importance of a unified view of the mathematical sciences, an approach that is appealing to today's students as well as inherently desirable. Despite the fact that most of the faculty were trained in pure mathematics, they all respect, and teach with enthusiasm, the different areas of the mathematical sciences. Many have returned to graduate school to get formal training in computer science and other applied areas. Furthermore, they are cognizant of the need to "sell" mathematics to students today.

The shortage of broadly trained mathematics graduates requires greater publicity. In 1970, 4.5% of entering freshman planned to be mathematics majors, while in 1980 fewer than 1% showed a similar preference. The recent upsurge in student interest in good careers and "professional" training has bypassed mathematics (but not computer science). In the 1960's, students looked to mathematics as a basic subject for any scientific career. Although more students are now taking calculus than ever before, these students frequently view calculus as a necessary evil required for, say, entrance to medical school. The successful mathematics programs teach calculus with applications to the social as well as to the physical sciences. They often do some numerical analysis (with computers) and mathematical modeling in the calculus courses. Faculty in these successful programs view all service mathematics courses—calculus, statistics, finite mathematics, even introduction to computing—as opportunities to show off the value of mathematics in the modern world. In personal, out-of-class conversations with calculus students, they talk about other interesting, useful mathematical subjects, such as differential equations or operations research, and the rewarding careers available to graduates with a broad training in pure and applied mathematics. They say little about graduate study in pure mathematics to freshman and sophomores; that comes later, and only to students who show an inclination towards such study.

In 1970, 4.5% *of entering freshman planned to be mathematics majors, while in* 1980 *fewer than* 1% *showed a similar preference.*

Students are in a sense forcing faculty to sell mathematics, perhaps wrongly because the students are a bit lazy and wary to undertake a difficult subject like mathematics, perhaps rightly because they do not unquestioningly obey when told that math is good for them. In any case, I believe that the effect on faculty is good, forcing the faculty to view prospective careers in mathematics from the students' point of view and making faculty see how much more attractive the whole subject becomes with a unified approach to the mathematical sciences.

I believe that the mathematics major must become a broader mathematical sciences major, built on a combination of the values of abstraction and analytical reasoning central to pure mathematics and the modeling and applicable techniques of applied mathematics. Mathematics professors must be wary of approaching a mathematical sciences major with biases born from their own graduate training in pure mathematics. (Few applied mathematicians now go into college mathematics teaching, choosing instead positions in industry or in specialized university departments.) Having a department with narrowly oriented pure and applied mathematicians is no substitute for having all faculty practicing a unified approach to the mathematical sciences.

Once faculty have converted themselves to this unified thinking, they should find it easy to attract students into their mathematics programs. The one challenge they will then face is that many students lose interest in mathematics and develop sloppy mathematical thinking in high school or earlier. College faculty must now work also with high school colleagues to find ways to make high school students appreciate the value of mathematical skills in the modern world.

References

[1]   Committee on the Undergraduate Program in Mathematics. *A Compendium of CUPM Recommendations, Vols. I & II*. Mathematical Association of America, Washington, D.C., 1978.

[2]   ACM Curriculum Committee on Computer Science. "Curriculum 78: Recommendations for the Undergraduate Program in Computer Science." *Comm. Assoc. Comp. Mach.* 22 (1979) 148-166.

[3]   James T. Fey, Donald J. Albers and John Jewett. *Undergraduate Mathematical Sciences in Universities, Four-Year Colleges, and Two-Year Colleges, 1975-76*. Conference Board of the Mathematical Sciences, Washington, D.C., 1976.

[4]   W. Duren. "CUPM, The History of an Idea." *Amer. Math. Monthly* 74 (1967) 23-37.

Purity in Applications

Tim Poston

When I was seventeen I regarded myself as a Pure Mathematician. There was perhaps arrogance in this—mathematically my academic record was not prodigious, and my purity was open to all sorts of doubt—but it is worth considering what I meant by it.

First, it did not mean that of the three courses I was by then taking in the early specialized British system—"Pure Maths," "Applied Maths," and "Physics"—I preferred the first. The "Pure" syllabus was an intellectual ragbag. Some series, some simple ordinary differential equations, some standard integrals which I could memorize for just over the time of an exam, a smattering of the axiomatics of three-dimensional Euclidean geometry . . . as coherent as the dates of kings. By contrast the "Applied" course centered on a few rich concepts (force, energy, velocity, mass, momentum . . .) and a few wild fictions (the inextensible or perfectly elastic string, the rigid rod, the smooth wire) whose function was to exclude everything *but* those concepts from the universe of discourse. None of the messy irregularities of machines that might actually do something. More techniques relevant to the average user of mathematics were taught in the "Pure" course, but "Applied" had the majesty of a (small) bust of Newton through its coherence and sense of pattern. As a budding scientist I would have been disgusted by irrelevance: as a "Pure Mathematician" I basked in its beauty.

Tim Poston is currently a Professor at the Institut für Theoretische Physik of the University of Stuttgart. He received his first degree in mathematics from the University of Hull (England) in 1967, and began postgraduate work there. After a sabbatical year as student union president he transferred to the University of Warwick Mathematics Institute and completed his doctoral work in 1972. Since then he has worked at various universities and research institutes in Rio de Janeiro, Rochester (New York), Porto (Portugal), Geneva, and Bristol. His research has been generally concerned with catastrophe theory, ranging from the mathematical aspects through applications to plate buckling and archaeology. He has co-authored texts on differential geometry and relativity and on catastrophe theory and its applications.

I fully expected this pattern to change at University, as it did. The Department of Pure Mathematics offered, at last, theories that combined the fine workmanship of rigorous argument with the noble unity of the great cathedrals. I know I was still in the lesser chapels (the books still had "Elementary" on their backs) but the style was unmistakable.

The other Department was a negative miracle. The clean intellectual dissection which could take apart one of those improbable smooth wire problems, argue the way to an equation whose solution demanded few messy sums, and present an elegant answer, was gone. There was no more realism than at school, but the extra dimensions involved in electromagnetism and gyroscopes meant that clear argument at an elementary level was no longer enough.

It was not replaced by coherent reasoning at a more advanced level.

It was replaced by the debauch of indices.

I cannot conceive that the idea of a function defined on triples of geometric vectors, giving the volume of the parallelepiped with vectors u, v, w for edges, is more abstruse and advanced than the mysterious symbol ϵ_{ijk} defined as "$+1$ for i, j, k a cyclic permutation of $1, 2, 3$, -1 for a reversed permutation, and 0 if two of i, j, k coincide." Certainly ϵ_{ijk} represents the volume function, conveniently, in some coordinates. As an aid to calculation it is invaluable. But often a little clear thought can replace a great deal of calculation. Surely volume is an easier subject for clear thought than a set of 27 1's, 0's and (-1)'s which obscurely requires multiplication by a Jacobian determinant in some coordinate systems. (Again, this *need* not be obscure, if the idea of an operator A—represented by, but not consisting of, a matrix—has been made clear, so that det A can be clearly introduced as the number A multiplies volumes by.)

It is good to quote Heaviside's defense, in 1893 [2], of the use of mathematics in science:

> Facts are not much use, considered as facts. They bewilder by their number and their apparent incoherency. Let them be digested into theory, however, and brought into mutual harmony, and it is another matter. Theory is of the essence of facts. Without theory scientific knowledge would be only worthy of the mad house.

Within mathematics, the facts are the formulae. They must be digested, brought into mutual, intellectual harmony or they are only worthy of a minor Applied Mathematics Department. This digestion, this bringing into harmony, is "pure" mathematics as a musical sound or a voice is pure: "Free from roughness, harshness or discordant quality; smooth, clear" [Oxford English Dictionary]. It is also Pure Mathematics, in that it is not linked to any one particular application. The volume of a bucket, and the volume of a region in n-dimensional phase space, are both illuminated by thinking "in the abstract" about the concept of volume. The practical advantages of intellectual clarity can be enormous.

For example, in the very early (pre-electronic) days of computers, a computer group was given a large numerical matrix to invert by some aerodynamicists [1]. The inverse that a major effort produced, to several significant figures, was . . . the transpose. A brief examination of the original, abstract matrix (with functions for entries, not the evaluated numbers the computer group was first given) showed it to be orthogonal. The inverse could have been known to be *precisely* the transpose, with no heavy calculation whatever. True, it is now so cheap to invert a matrix that some people will argue that studying it intellectually first is not cost-effective; but that way lies the mad house of the compulsive programmer "with sunken glowing eyes" who Weizenbaum [6] says "has only technique, not knowledge." In a hurry, yes, press the button for an inverse. But if the habit of understanding is lost at an elementary level, or never learned, it will not reappear when the problems become more complicated.

Some years ago I became involved in numerical calculation of crystal spectra [4], where vast and intricate routines, cunning in all their details, took hours to crunch their way. Since I had no training in this area I was not buried in technique, and could observe that the basic approach was as perverse as the numerical inversion of orthogonal matrices. If the "dispersion relation" between frequency v and wave number k was treated not as giving a "branched function" $v_i(k)$ but as a dependence of one component of k on v and the others, costly root searches could often be replaced by simple evaluation of functions. In any case a far more general, flexible and cheap approach resulted.

If the habit of understanding is lost at an elementary level, or never learned, it will not reappear when the problems become more complicated.

Conceptual thinking is the salt of mathematics. If the salt has lost its savour, with what shall applications be salted? In training users of mathematics, the question of what topics—viewed as techniques—from "Pure Mathematics" should be taught them is often a matter of fierce debate. But it is not fundamental. More important is the pure essence of mathematics, the understanding of structure, and intellectual harmony. If a student learns nothing of this, he or she has learned no mathematics; and the scientist, economist or engineer unconscious that mathematical understanding offers dollars-and-rubles rewards is half blind—even if the other eye sees grandly. Heaviside [2], again:

> Unfortunately, in my opinion, Faraday was not a mathematician. It can scarcely be doubted that had he been one, he would have been greatly assisted in his researches, have saved himself much useless speculation, and would have anticipated much later work . . . But it is perhaps too much to expect a man to be both the prince of experimentalists and a competent mathematician.

However, it should be reasonable to expect a man to be both an Applied Mathematics specialist and a competent mathematician in the sense developed above, with some skill in seeing the wood as well as the trees.

Sometimes, this expectation is justified.

Far too often, it is not. In Britain particularly, history has produced a whole tribe who are not mathematicians in this sense—and are not really working on applications either. Beads on wires have their higher analogues, largely drawn from 19th century continuum mechanics. General relativity, however, has found a place, perhaps because in classical notation it has so many indexed quantities—though not, usually, general relativity as expounded in the beautiful book of Misner, Thorne and Wheeler [3]. That book is addressed to people with a real interest in applications, like physicists and astronomers.

Calling the pseudo-Riemannian manifolds involved "spacetimes" made the work Applied, though not applied to the only universe in town.

More representative of the bad pattern I am discussing (which is *not* universal, but is widespread) was a semester's work by a graduate student I knew, on whether one high-dimensional "spacetime" would embed in another. Calling the pseudo-Riemannian manifolds involved "spacetimes" made the work Applied, though not applied to the only universe in town. High-dimensional embedding problems are meat and drink to modern differential geometry, of course, which has developed a battery of powerful conceptual techniques for their study. Did he seek to learn these? Did he, even better, try to develop his own conceptual tools from scratch? (Zeeman [7] gives an excellent example of the merits of working from ignorance in a similar case.)

He did not. He got stuck into the equations in their indexed glory, and ground through. At the end he thought he had a positive result, particular rather than general of course, but that was what he had been asked for. Then, he checked. Two weeks in, he had made a slip in the calculations, invalidating the manipulations of the rest of the semester. The same time spent on conceptual work (with calculations where needed) would have greatly enlarged his understanding of pseudo-Riemannian manifolds, even if nothing publishable resulted. As it was, he had gained only another slab of practice at manipulating indexed-quantity equations.

My claim is not, of course, that nothing without immediate application should be studied. Apart from the non-application virtues of mathematics (compelling, but hard to establish with a listener who *needs* an argument), mathematically natural questions usually do tie up with applications sooner or later, as the algebraists' invention of complex numbers took over electronics and quantum mechanics. Pseudo-Riemannian structures are not just a quirk of the physicists: there is a natural one on, for instance, the group of area-preserving linear tranformations of the plane. They would surely have been invented, and almost as surely been applied, even if relativity had not come along, and in an application other than to physical spacetime the dimension might be anything. But in the absence, so far, of such an application, the above work was as "Pure" as anything done in a Pure Mathematics Department, failing only to be pure, and perhaps to be mathematics. (This last depends on definition. Does an IBM 704 do mathematics?)

Not all "Pure Mathematics" is pure either, in this sense. "The future does not lie with the theory of generalised left pseudo-heaps" [5], but a look through the journals shows that a substantial part of the present lies with just such forgettable material. Academic Pure Mathematics has vices all its own, but they are not my subject here, which is purity in application.

Purity, it seems to me, is more fundamental a criterion than mathematical rigour. Indeed, rigour is basically a tool for purity, for penetrating to an insight as to just what is really going on. Historically, its meaning has varied, and the notion of a correct proof has usually changed in response to a crises in understanding. When natural arguments could sum $1 - 1 + 1 - 1 + \ldots$ as 0, 1/2 or 1, it became necessary to sharpen the reasoning tools used. Hilbert foresaw a climax to this process of refinement, but Gödel gave us leave to doubt the finality of all arithmetic arguments, forever. In any case, even Bourbaki "abuses language" at the expense of rigour, to benefit purity. In applications a demand for rigour, as an end in itself and not as a servant of purity, can be even more deadening than the debauch of indices. It can even encourage the growth of bogus "Applied" theories, like the equations of irrotational inviscid flow von Neumann called the study of dry water, by masking the assumptions made in a wreath of technicalities. A crowded map showing every logical leaf of every demonstrational tree may so effectively conceal the wood that nobody will notice that the thing being mapped is not a forest at all but grassland. Among a list of fifty-three axioms about the cell structure of the plants in the model, who will notice the importance of the presence of lignin?

Rigour has a crucial part to play in clarification. For example the question "In what space do these objects you are calculating with live?" is an appeal for rigour that sheds light from classical electromagnetism to wave mechanics to the buckling of steel plates. Sometimes the questions raised by a wish for purity cannot be quickly answered. Poincaré's creation of topology out of his purity in mechanics raised so many pure mathematical questions that the subject went into purdah for half a century, and has only recently started acquiring its natural prominence in physics. Approximations by Taylor expansions were in use centuries before the techniques associated with catastrophe theory gave lucid, rigorous reasons behind their perfect validity for some purposes, in suitable cases, and proved their inadequacy in others. Purity is a goal and a standard that can never be perfectly met, and must (like rigour) often be compromised for the sake of some numerical answer demanded by a more immediate goal. But the compromise should be conscious, and considered: for if the surrender of purity is complete, the result is first computational ineffectiveness and finally the mad house.

I no longer call myself a Pure Mathematician—indeed, I am currently in a Physics Department. I do not want the name Applied Mathematician, with its British associations and self-exposing illogic (who applies the mathematician?). I do not work in one, single field of science, whose name I could borrow; I have published with coauthors from physics, geography, archaeology . . . I am still a mathematician. I work, as best I can, in the application of mathematical understanding. To the extent that I succeed, in the meaning here proposed I am a pure mathematician.

References

[1] Forman S. Acton. *Numerical Methods that Work*. Harper and Row, New York, 1970.
[2] Oliver Heaviside. *Electromagnetic Theory*. D. Van Nostrand, New York; and "The Electrician," Printing and Publishing Company, London, 1893.
[3] C.W. Misner, K.S. Thorne, J.A. Wheeler. *Gravitation*. W.H. Freeman, San Francisco, 1973.
[4] T. Poston and A.B. Budgor. "A Geometrical Approach to Calculating the Energy and Frequency Spectra of Crystals." *J. Comp. Phys.* 19 (1975) 1-28.
[5] I.N. Stewart. *Concepts of Modern Mathematics*. Penguin, London, 1977.
[6] Joseph Weizenbaum. *Computer Power and Human Reason*. W.H. Freeman, San Francisco, 1976.
[7] E.C. Zeeman. "Research Ancient and Modern." *Bull. IMA* 10 (1974) 272-281; also in E.C. Zeeman. *Catastrophe Theory: Selected Papers* 1972-77. Addison-Wesley, Reading, Massachusetts, 1977.

Growth and New Intuitions:
Can We Meet the Challenge?

William F. Lucas

The mathematical sciences have changed significantly during the past few decades. The most obvious change is the enormous growth of mathematics. However, the most exciting and potentially beneficial movement may well be the extensive mathematization of many traditional as well as newly emerging disciplines. The consequent influx of rich new ideas and alternative intuitional sources can greatly rejuvenate and invigorate mathematics itself. It would thus appear that mathematics should, for some time into the future, exhibit a truly great scientific advance that would bring reasonable prosperity to both its individual practitioners and its supporting institutions. Nevertheless, there is ample evidence to indicate that these possibilities are not being realized, and the prospects for the future are much less encouraging. The most obvious illustration of this is the gross mismatch between the mathematics that students are currently being taught and the skills that are marketable to most current users of the subject.

The huge gap between the great potential for mathematics and the traumatic events of the "real world" must have some logical basis and root causes. Some bad decisions were made in the past, and current responses appear dreadfully inadequate. To ignore these problems by continuing in our current mode of operation (or inaction) will do a great disservice to our

William F. Lucas is Professor of Applied Mathematics and Operations Research at Cornell University. He was director of Cornell's Center for Applied Mathematics from 1971-74, following teaching and research appointments at Princeton University, the Rand Corporation, the Middle East Technical University in Ankara, and the University of Wisconsin. He received his Ph.D. in mathematics from the University of Michigan in 1963, following masters degrees (in physics and in mathematics) and a B.S. from the University of Detroit. Lucas has served as chairman of the MAA's Committee on the Undergraduate Program in Mathematics and of the Committee on Institutes and Workshops. He has directed or lectured in many short courses on contemporary applications of mathematics, including the AAAS Chautauqua-type short courses. He is known for his research in game theory and is currently scientific editor of the *International Journal of Game Theory*.

subject, to those who are or will be involved in it, as well as to a large element of society which increasingly uses our product and which is significantly affected by the nature and consequences of mathematical developments.

The commonly held view that mathematical directions are somehow independent of the stresses acting on them from the outside world is a naive one. Clearly, mathematics has been and will be profoundly affected by external events. A recent dramatic example is the fact that modern technology developed the electronic *digital* computer, rather than an analog machine. Mathematics will not remain a vital area for long if many of its students cannot find suitable employment. It is certainly time for the mathematical community to recognize and face up to reality, to rationally analyze the situation with an open mind and from a broad perspective, to follow the evidence wherever it leads, to institute new leadership when necessary, and to summon the fortitude to undertake the corrective measures which should prove best for all involved, and not merely for the few already profiting from the present situation.

There will be little room for elitism, for prejudice, or for the "Hardy syndrome" (the less useful the better). Significant changes in our current attitudes and in typical "respect curves" will be necessary. (One usually has a high level of respect for fields rather close to his or her own specialty, as well as for those very far from it, whereas respect for disciplines at more intermediate distances is often much less. For example, a mathematician may hold the physical sciences in high regard and have esteem for the humanities, but have little respect for business or the social sciences.)

In this paper the term "mathematics" will be used in the very broad sense of the general mathematical sciences and its applications. It seems best, in order to avoid confusion, to place the adjectives where they belong: "pure" before mathematics or "classical" before applied. After all, mathematics is a subject rather devoid of adjectives. Moreover, it is best to avoid misconceptions of the type in which one who strongly supports "applied mathematics" means only that mathematics commonly used in the physical sciences, and may really be just as elitist about this area within applied mathematics as many mathematicians are about pure mathematics. Anyone who cannot distinguish between "applied" mathematics, properly defined, and "bad" mathematics should no longer be taken seriously.

Growth and new ideas

The most conspicious and impressive development in mathematics is the enormous increase in the amount of new subject matter being created. The quantity of new knowledge is doubling about every decade. Evidence of this can be seen in the increase in research publications, in new books and journals, in the size of *Mathematical Reviews*, in the numbers of new Ph.D.'s and practitioners of the subject, and in the vast penetration of mathematics into other fields. Some may contend that a good deal of low quality work is published in this day of "publish or perish," or that most new discoveries are so advanced or specialized as to not be relevant to users or educators. On the other hand, a close examination of the older mathematical literature will likely convince one that the fraction of poor quality publications was probably as high in the "old days" as it is now. One could further argue, with so many new fields and applications being created within mathematics, that a good portion of these newer areas are more accessible to a large number of researchers and educators than are many of the more developed parts of traditional mathematics. In fact the research frontiers in newer subjects are often closer in terms of prerequisites, although not necessarily easier to push back. In addition, this growth in mathematics is taking place in both pure and applied directions, uses continuous and discrete techniques, covers traditional and new areas, and often appears in conjunction with advances in a large number of other disciplines.

This spectacular growth of mathematics is rather contrary to the layman's view of this field, and is probably not fully appreciated even by most teachers of the subject. The greatest part of the typical educated person's own mathematical education has been a narrow exposure to rather classical and traditional topics. Rarely (outside of the computer area) have such persons learned of anything originating in the past 20 years, or even much in the last 200 years, and in some cases little from the past 2,000 years. Although it has been improving recently, there is still extremely little coverage of mathematical topics in the more popular news media. Mathematicians have few well-known heroes and no Nobel prize winners to capture public attention. This drab image contrasts sharply with the dynamic, creative reputation of neighboring fields such as the physical and life sciences, and much of engineering. Worse still, the same dull image portrayed by the popular press and electronic media frequently dominates the classroom as well. Many benefits would clearly follow if the mathematical sciences could achieve a greater visibility as a vital contemporary field.

Most mathematical creations are the result of intuition, a rather direct perception of fact or rapid apprehension rather independent of any long or formal reasoning process. Many such insights, however, do follow long observation or thorough understanding of some physical activity or other concrete mental construction. Often mathematical theories begin from simple observations about numbers, space, physical changes, or engineering devices. The direction of modern mathematics has without a doubt been greatly influenced by developments in other disciplines. One could argue that one of the most revolutionary things going on in mathematics today is the great increase in new subjects that can serve as sources for new intuitions and creativity in mathematical science. Some of these sources arise within mathematics itself. On the other hand, many other disciplines are being seriously scrutinized from a mathematical point of view for the first time in this generation. These include large parts of the social, behavioral, managerial, decision and life sciences; they lead to quantitative considerations of such human and social concerns as utility, measurement, bargaining, equity, conflict, human values, and many others. It is safe to say that developments in these directions (as those involving digital computers and discrete mathematics) will have revolutionary impact on mathematics, even though they are currently only in an early stage of development. It is also rather clear that radically different mathematical theories must be created in order to assist these other fields; it will not be enough to merely repackage ideas that arose out of physics or geometry. Interactions with those newly mathematicized disciplines should both influence significantly the growth of mathematics, as well as broaden the scope of employment opportunities for those trained in mathematics.

John von Neumann's well known commentary on the relevance for mathematics of empirical sources is worth repeating today, 30 years after he first expressed it:

> As a mathematical discipline travels far from its empirical source, or still more, if it is a second and third generation only indirectly inspired by ideas coming from "reality," it is beset with very grave dangers. It becomes more and more purely aestheticizing, more and more purely *l'art pour l'art*. This need not be bad, if the field is surrounded by correlated subjects which still have closer empirical connections, or if the discipline is under the influence of men with an exceptionally well-developed taste. But there is a grave danger that the subject will develop along the line of least resistance, that the stream, so far from its source, will separate into a multitude of insignificant branches, and that the discipline will become a disorganized mass of details and complexities. In other words, at a great distance from its empirical source, or after much "abstract" imbreeding, a mathematical subject is in danger of degeneration. At the inception the style is usually classical; when it shows signs of becoming baroque, then the danger signal is up. . . . In any event, whenever this stage is reached, the only remedy seems to me to be the rejuvenating return to the source: the reinjection of more or less directly empirical ideas. I am convinced that this was a necessary condition to conserve the freshness and the vitality of the subject and that this will remain equally true in the future.

Similar views were often expressed by Hermann Weyl and Richard Courant, individuals who, like von Neumann, excelled at both pure and applied mathematics. A personal friend of von Neumann has told me that von Neumann often expressed a very deep related concern about the rather haphazard way mathematical specialties were selected to receive the limited resources of human talent and financial support.

Problems in education

It has become increasingly clear that there are very serious problems at all levels of mathematics education in the United States. One can place the blame for many of these difficulties on changes and attitudes within the larger educational establishment or in the general society, and thus beyond the control of mathematicians. However, this is taking the easy way out. For many of these problems have been brought about rather directly as a result of well-intended but misguided decisions made by mathematicians and mathematical educators.

Most people now accept as fact that the "new math" has been a failure in the elementary schools, although many individuals and groups who supported it in the past are still attempting to defend it. The idea that one should teach young children mathematics in a framework of set theory, axioms, logic and sophisticated vocabulary was probably not a very good

---

It is sad to see . . . the rush to "return to basics," since they were not all that exciting or successful . . . there are so many other new idioms which could be used as a matrix in which to embed elementary mathematics.

---

one to begin with, and the way it was implemented was sure to doom this approach. (There are many interesting reports which debate this topic; see, for example, [20] and [21].) It is sad to see the public response, the rush to "return to basics," since they were not all that exciting or successful either. This return to a nineteenth century curriculum is especially unfortunate when there are so many other new idioms which could be used as a matrix in which to embed elementary mathematics. (Many interesting elementary examples could be selected from the recent flood of advances in finite mathematics.) Unfortunately, educators appear too timid to experiment with a "new new math" in light of the failure of the old "new math," and are often not well enough informed of contemporary developments to select a new and appropriate approach.

Many now agree that the low point in U.S. education is in the junior high schools. The list of possible reasons is a very long one. The philosophy of behavioral objectives, self paced or individualized instruction, the non-prerequisite nature of many new teaching materials (and resulting increases in short-term memory), as well as many other experimental approaches severely hurt mathematics education. The most glaring evidence that all is not well in the high schools is the enormous remedial problem the colleges now have with incoming freshmen. Moreover, the serious problem of finding a sufficient number of high school teachers knowledgeable about calculus, computers, microprocessors, statistics, and contemporary applications of mathematics will probably prevent a proper response by most school systems.

The current and projected employment situation for those with degrees in mathematics at the college and graduate levels surely indicates that the content of most current educational programs is badly "out-of-sync" with the employers' hiring practices. One could introduce a lengthy discussion at this point about the goals and purposes of a liberal arts education, but the interested reader could reread Cardinal John Henry Newman or other good sources instead. Surely there is a wide middle ground between a "culturally rich" education deemed suitable for the college graduate of the 1920's (or 1950's if you prefer), and a training aimed only at earning a good living. Society values those who can think logically, communicate well in a concise manner, and follow a lengthy chain of thought, as well as those with tradeschool-type skills. In any event, graduate education has never been free to ignore the realities of the job market. Many proposed graduate programs do in fact get "shot down" on grounds that jobs for its graduates may not exist. And some existing programs will likely be terminated for this same reason! The main point, which is made very well by Lotfi A. Zadeh in an article entitled "Mathematics: A call for reorientation" [3], is that mathematical education at the higher level is in a crises situation. Many people believe that major changes are drastically needed.

Counterexamples to good advice

It is now clear that many judgments and recommendations made by past leaders of the mathematical community have been less than good, if not outright disasterous. It is thus reasonable to consider the possibility that many current suggestions and proposed directions originating from the same general sources might also be highly questionable, if not downright ill-advised. Though many others could be cited, only a few of these, to serve as illustrations, will be touched on here.

Perhaps the most notorious particular case is the excessive need for people with advanced training in mathematics which was predicted in the rather self-serving COSRIMS report in 1968 [13]. Whereas some social scientists had already provided a much more accurate picture of what was to occur in various disciplines, and published such in popular periodicals a few years earlier. Although the COSRIMS volumes contain a great deal of valuable information, their conclusions on this point serve as a living example of the need for more and better "applied mathematics" within the mathematical community itself. Some have stated, in defense of COS-RIMS, that their conclusions about future needs were correct logical deductions from the assumptions put into their model. But this only serves to make the point that those dealing with the real world cannot afford the luxury of starting off with just any old set of consistent axioms.

Probably the greatest damage to the welfare of mathematics and its practitioners was that caused by the elitist attitudes of mathematicians between about 1950 and 1975 which led to the essential exclusion from most mathematics departments of huge new areas such as computer science, operations research, combinatorics, and mathematical social science. As is frequently the case when such discrimination is practiced, the mathematicians fenced themselves in rather than holding back the newcomers or minority groups of the time. In reference to just theoretical computer science, Zadeh refers to this exclusion as "a non-event that could have altered the course of mathematics in a fundamental way." One naturally wonders what can be done now to undo the harm caused by such discrimination in the past, or more importantly, what similar mistakes (often caused by nondecisions, inactions or ignorance) are currently being made.

Probably the greatest damage to the welfare of mathematics ... was that caused by the elitist attitudes of mathematicians ... which led to the essential exclusion from most mathematics departments of huge new areas such as computer science, operations research, combinatorics, and mathematical social science.

The great growth in the annual number of new Ph.D.'s between the mid-1960's and the mid-1970's was certainly good for the vitality of mathematical research. It is reasonable to assume that as this number grew, the much smaller number of really outstanding new young scientists who would make truly major research contributions to mathematics was also increasing, perhaps even in a roughly proportional manner. It seems equally reasonable to assume that as this total number of new Ph.D.'s decreases, that the small percent in the latter distinguished category (who are perhaps less likely to need to worry about employment) will likewise decrease. It also seems quite obvious that some of the spokesmen for our professional societies wanted to maintain the high number of new Ph.D.'s in this top category while significantly decreasing the total number. This is illustrated by the open letter of November 11, 1975 by Lipman Bers that appeared on the inside cover of the *Notices of the AMS* in conjunction with statements by various officials at panel discussions at AMS meetings. Nevertheless, more recent evidence seems to indicate that we are losing the top quality graduate students at a rate just as fast, or even faster than, the total number. This is hardly a surprise. The really creative college senior is often the one with the most flexibility, confidence and courage. Such a person can choose graduate programs in other exciting areas such as computer science, electrical engineering, physical science, or economics. Why should he or she pursue an area which advertises its difficulties? The undergraduate who is capable of pushing back the frontiers of mathematics in future years is hardly going to join an area in recession just "for the love of it" when there are so many other prosperous and exciting subjects to pursue. Mathematics itself might be much better off if our leaders put more serious efforts into really altering the nature of the new Ph.D.'s, thereby increasing job opportunities, rather than discouraging young people from the area.

In recent years we have been advised that any Ph.D. in mathematics with the right attitudes and intentions can teach nearly any undergraduate mathematics course. (Some small departments have even experimented with faculties primarily in one area for the advantages it provides for research.) For the good of mathematics and its best young researchers, it would thus seem reasonable to follow a policy of hiring the best scientists for most undergraduate teaching positions independent of their specialties. According to this reasoning, mathematics departments can continue to hire experts in topology and algebra even though the courses they need to cover are in numerical analysis, statistics, operations research, mathematical economics or combinatorics. Of course, this supposition is symmetrical: departments should also hire applicants in differential equations, mechanics, or relativity theory when they are the best ones, even if the undergraduate courses to be covered are in algebra, analysis, topology or logic. I dare say that there are many counterexamples to demonstrate that this policy, in the latter direction, was not in fact practiced much in the 1960's. At that time people seemed well able to distinguish an algebraic topologist from a fluid dynamicist, and hired accordingly. Do we really believe that in areas dealing with contemporary applications which are changing so rapidly, even at the elementary levels, that the students are best served by having these courses taught by pure mathematicians trained in the more fashionable or traditional areas rather than by qualified specialists? Many people believe just the opposite, that in order to improve offerings in new applied directions the best thing to do is to simply hire capable people in these fields. To maintain the status quo of faculty distribution will hardly create flourishing programs in new, useful applications, nor fool the students into flocking to upper division mathematics courses.

Misuse of axioms

Some now feel that the most disastrous thing to hit mathematics education in the past generation has been the premature introduction and excessive use of the axiomatic method. Whether subjugation to the "tyranny of logic" stifles mathematical creativity at more advanced levels is debatable. Nevertheless, it is much more obvious that the emphasis of this approach is less than optimal for many educational purposes. The failure of the "new math" at the elementary level and the remedial problems arising at the secondary level give ample evidence of this. L.C. Wood's article [12] "Beware of axiomatics in applied mathematics" warns of the danger of excessive use of the axiomatic method in classical applied mathematics and its misuse in teaching the physical sciences. In a recent interview [1] about the traditional field of geometry, H.S.M. Coxeter spoke of

> ... dull teaching: perhaps too much emphasis on axiomatics went on for too long a time. ... So children got bogged down in this formal stuff and didn't get a lively feel for this subject. That did a lot of harm.

It is also rather clear that the stress on using axioms is often put in the wrong place. Consider, for example, the role of axioms in applied mathematics. As the custodian cleaning up the workshop after the inventor has gone home, axioms in the physical sciences usually appear towards the end of a long chain of creative activities, when one finally refines a theory to precise, concise, and compact form. Frequently this is done in order to

---

Axioms are often a very integral part of the early modeling process in theoretical approaches to many problems in the social, decision and managerial sciences.

---

teach it more rapidly (so that one can perhaps get on more quickly to recent results). In contrast, axioms are often a very integral part of the early modeling process in theoretical approaches to many problems in the social, decision and managerial sciences; popular topics such as social choice theory provide many examples of simple, effective axiom systems. Basic principals are quickly introduced as axioms, and questions of existence and uniqueness arise immediately in a very natural way. Unfortunately, such interesting, creative and necessary uses of axioms typically receive very little attention in our classes, while a great amount of time is spent on using axioms to convince students of obvious propositions about common number systems or plane geometry, or to quickly explain physical phenomena without going through all the experimental or empirical steps which took place historically.

The greatest villains in the mathematical community are probably those with the short memories who have been responsible for taking the "fun" out of mathematics. There are surely a great number of fundamental and yet much more exciting or stimulating things to be done in mathematics than the "new math" or premature emphasis on abstraction *per se.* Many superb students in the advanced tracks in high school or lower division courses in college who indicate an interest in pursuing mathematics seem to switch to other fields before they find mathematics overly challenging; frequently they switch because their interest and inspiration were not maintained, rather than because of some newly discovered love for another field. Advanced graduate students in mathematics will frequently admit that if they had known in advance what the emphasis would be in their courses, they would have more seriously considered alternate subjects.

Teachers of mathematics would do well to consider whether their teaching emphasizes what they and their colleagues found most exciting or influential in selecting their own careers. Many different aspects of mathematics would likely get mentioned as "fun" or as most inspirational. Yet one can conjecture that the results of such a survey would differ significantly from current teaching emphases. Of course, some reasonable trade-off is needed between recapturing the most popular, and including what is "best" for the student. Current offerings do not, however, appear to be the most desirable compromise. One way to cut down on excessive abstraction is to avoid its use when the students are not familiar with at least two concrete realizations of the thing being abstracted. We may have to worry less about such fads as "math anxiety" or "math avoidance" if we put a little more "fun" back into our courses.

Changes in education

The great growth in sheer volume of new mathematics as well as its many new uses is probably the main cause of many of the stresses arising in college and graduate level education. An undergraduate mathematics major without a couple of courses in computer science and probability and statistics is viewed as illiterate by a large number of employers. It is truly sad to see current mathematics majors with so little appreciation of the extensive and essential uses of classical applied mathematics and the great success story achieved by it in the physical sciences. (See the writings of Morris Kline for more on this, but read "classical applied mathematics" where he uses "applied mathematics.") Furthermore, many mathematics majors seem rather dated and deprived (and thus less employable) when they do not have some familiarity with modern developments in operations research, in the uses of mathematics in the social or managerial sciences, as well as some knowledge of the ongoing revolution in finite mathematics. How can anyone expect to cover most of these topics well while at the same time including calculus and those upper division "core" courses containing what every mathematics major should know? How can any student absorb, sort out, and incorporate the valuable new developments so rapidly?

With the (negative) "half life" of mathematics being ten years or less, one could certainly make a strong argument for the inclusion of much more recently discovered mathematics in addition to or in place of the disproportionate amount of old material currently covered in the standard curriculum. It does not seem unreasonable to propose to mathematics teachers that they seriously ask themselves the following sequence of questions.

What percent of the material in my courses was newly discovered or mostly developed in the past $10n$ years? If $n = 1$, one is talking about the most recent *half* of mathematics! If $n = 4$, and one's answer is "only a small percent," then this person is excluding most of mathematics! If a teacher is not embarrassed or concerned with the fact that he or she teaches a disproportionate amount of "older stuff," then educators in general or the

> *Although some older art, music or wines may be better than the newer, it is rather unlikely that this would often apply to science or mathematics.*

public at large should be entitled to a good explanation of why this is the case. Although some older art, music or wines may be better than the newer, it is rather unlikely that this would often apply to science or mathematics. It seems unlikely that the older parts of such a discipline really are somehow the most basic, the best "culturally," the real "core" of the subject, the ones with the holding power to "win" in the long run (in light of their holding power to date), the most beautiful intrinsically, or the most useful within or outside of the subject. A college (or even a high school) course in physics or biology (not to mention astronomy, geology or genetics) which contained as little of the ideas developed in the past 25 years as most of our mathematics courses would be considered obsolete.

One possible answer is to increase the amount of mathematics taken at the B.S. and M.S. levels. This may seem less objectionable to those favoring a broad liberal education if we put across the idea of a "division" of mathematical sciences similar to what is the common view of the physical, social or life sciences, in contrast to the current image of mathematics as a single "department." Alternatively, perhaps calculus and computer science courses should not count toward the required number of hours for the mathematics major, as is the case for physics, chemistry and engineering majors. After all, few mathematics majors now minor (or earn a second major) in physics as was common in the past. In many schools the number of hours for a mathematics major is much less than the number for the accredited chemistry or biology majors, and in view of the labs the gap in contact hours may reach nearly 100%. A recent study in the College of Arts and Sciences at Cornell University proposed that students be allowed to count only 60 of their 120 hours for graduation from their major department. Serious objections to this limitation were raised by those in some of the laboratory sciences. On the other hand, altering structures and playing with the number of credit hours is only a small step in the direction of what must be done.

A more obvious, but highly controversial, approach is to re-evaluate the notion of the "core" and either shrink it in size or, if you prefer, accept the notion of different possible cores. For example, the new core for the mathematics major might consist of only one year of calculus, one semester of linear algebra, and one semester of real analysis. For purposes of argument, we should even consider the extreme case that the new core might be the empty set. This might now sound unacceptable, but it is very possible that it will occur in the next ten years. We may well see Ph.D.'s

---

The new core for the mathematics major might consist of only one year of calculus, one semester of linear algebra, and one semester of real analysis. . . . we should even consider the extreme case that the new core might be the empty set.

---

knowing only finite mathematics with degrees given by computer science or business programs, if not in mathematics departments. The methods of Donald Greenspan (which may well have been pursued earlier by such giants as Leibnitz and von Neumann if they had had modern day computers) suggest already how finite techniques and the computer can replace much of what is done in continuous applied mathematics. Additional ideas about alternate routes and different types of acceptable mathematics majors may appear in forthcoming reports from the Committee on the Undergraduate Program in Mathematics (CUPM) of the Mathematical Association of America (MAA). Excessive concern with premature selections of specialties (and those ever-influential "transfer students") should not delay the implementation of broader mathematical degree programs. Students have already made more crucial choices.

Another major problem must also be addressed. Where will the well qualified teachers for the new "new math" at the college and master's level come from? In a period of slow growth, a massive retraining or redirection will be necessary. It has not been uncommon to blame the failure of the old "new math" at the elementary level upon the existing teachers who had "old math minds" and attempted to explain the new language in terms of the old one it was replacing. It would be most unpopular if someone were to dare suggest that we now have the analogous problem at the college level. Nevertheless, a great deal of updating and reorientation will be necessary if our college faculties in mathematics are to successfully deal with the challenges of the 1980's and 1990's.

It might also be possible to be much more efficient in the teaching of mathematics. One does not have to be too observant to notice that most students in a typical mathematics class are not learning much, if anything, during most of the lectures. Questions about such material in future semesters often reveals little retention has occurred for material previously learned in or outside the classroom. Besides, mathematics teachers are usually not renowned for their studies or knowledge about the nature or rate of learning of mathematical concepts. They rely on personal experience! Many do agree that there are a variety of different approaches which are best for different students, and that the way a particular individual grasps mathematical ideas is often highly personalized. This suggests that mathematics classes should be much more interactive than is typical of traditional mathematics lectures. More emphasis on student involvement, various sorts of problem-solving courses, modeling courses, more intuitive applied courses, the R.L. Moore approach, team projects, undergraduate research opportunities (not only for the "best" students), and other approaches, if well integrated, may prove to be much more effective in learning well that which we value most in a mathematics education. Additional time for breadth of coverage could possibly result. (For more about interaction in mathematics courses and types of appropriate materials, see [19].)

The threat of inaction

The likely consequences of wrong or insufficient action on the part of the mathematical establishment in the U.S.A. are rather clear. The existing primary mathematics groups in this country will shrink in size and stature. Delays in initiating needed reforms of a substantial nature will lead to diminished vitality and relevance. In a prolonged period of no overall growth, little mobility and the unfortunate "constant-sum thinking" which currently so penetrates college administrations, it is highly questionable whether the mathematics community will be willing to undertake the necessary changes or to come up with the courageous leadership required to break out of its rather stagnant position and be one of the few groups to achieve positive growth. When the competition does heat up, others will begin to move (or push) in (especially at large institutions where several other groups are staffed with mathematical scientists) to fill the emerging vacuum left by mathematics groups. Many departments in fields such as electrical, industrial, systems and civil engineering, applied physics, computer science, and (in a "softer" way) business are already graduating some people who are essentially mathematics majors, at least towards the directions of modern applications. Anyone who has not recently examined such degree requirements and typical programs really should do so.

Simple symbolic devices such as altering a name from mathematics to mathematical sciences will hardly protect our frontiers from those areas standing in the wings waiting to replace us. And if the overall enrollment decline in higher education reaches a point of serious cuts in departmental sizes, then many other groups will decide that they too can teach their own mathematics service courses. (A possible scenario of this sort is suggested by a well-known but anonymous mathematician on pp. 42-43 of the

Proceedings of the MAA PRIME-80 Conference [18].) Some "alarmists" have gone so far as to say that mathematics departments will go the way of classical language departments. However, this is perhaps a bit extreme and is over-extending the metaphor of mathematics as a language. On the other hand, the author does consider a comparison with what has happened to philosophy departments in the United States as a realistic one. The latter subject was once viewed as relevant to the humanities (in particular, to history) as well as to theology (and thus studied in thousands of monasteries), just as mathematics has been considered essential to the physical sciences and engineering. The relevance of these subjects will not diminish, but the new and traditional user departments may well fight to teach their own *more relevant* courses in these basic subjects arguing, for example, the benefit of an integrated approach.

On the other hand, many golden opportunities still exist, despite our poor track record in recent years and the current press of time. The choice is up to the mathematics community, but it must act quickly and in a meaningful way. There exists many good strategies. The question is whether we will select one and implement it in time, or whether we will follow philosophy's decline into prestigious isolation and irrelevance.

References

[1]   Dave Logothetti. "An Interview with H.S.M. Coxeter, the King of Geometry." *The Two-Year College Mathematics Journal* 11 (1980) 2-19.

[2]   John von Neumann. "The Mathematician." In *The Works of the Mind*. Ed: R.B. Haywood. U. of Chicago Press, 1943.

[3]   Lotfi A. Zadeh. "Mathematics—A Call for Reorientation." *Newsletter, Conference Board of the Mathematical Sciences* 7 (May 1972) 1-3.

[4]   E.H. Bareiss. "The College Education for a Mathematician in Industry." *American Mathematical Monthly* 79 (1972) 972-984.

[5]   F.J. Dyson. "Missed Opportunities." *Bulletin of the American Mathematical Society* 78 (1972) 635-652.

[6]   Donald Greenspan. *Arithmetic Applied Mathematics*. Pergamon Press, 1980.

[7]   Saunders MacLane. "Mathematics." *Science* 209 (1980) 104-110.

[8]   Fred S. Roberts. "Is Calculus Necessary?" In *Proceedings of the Fourth International Congress on Mathematical Education*, 1980 (to appear).

[9]   Lester C. Thurow. *The Zero-sum Society: Distribution and the Possibilities for Economic Change*, MIT Press, 1980.

[10] Eugene P. Wigner. "The Unreasonable Effectiveness of Mathematics in the Natural Sciences." *Comm. Pure Appl. Math.* 13 (1960) 1-14.

[11] A.B. Willcox. "England Was Lost on the Playing Fields of Eton: A Parable for Mathematics." *American Mathematical Monthly* 80 (1973) 25-40.

[12] L.C. Wood. "Beware of Axiomatics in Applied Mathematics." *Bulletin of the Institute of Mathematics and Its Applications* 9 (1973) 41-44.

[13] *The Mathematical Sciences: A Report; Undergraduate Education*; and *A Collection of Essays*, U.S. National Academy of Sciences, 1968.

[14] "Education in Applied Mathematics." *SIAM Review* 9 (1967) 289-415; 15 (1973) 447-584.

[15] *Quarterly in Applied Mathematics* 30 (1972); Special issue on the future of applied mathematics.

[16] *The Role of Applications in the Undergraduate Mathematics Curriculum*, U.S. National Academy of Sciences (1979).

[17] *Mathematical Modeling: An International Journal*, Pergamon Press.

[18] *PRIME-80, Proceedings of a Conference on Prospects in Mathematics Education in the 1980's*, M.A.A. (1979).

[19] W.F. Lucas. "Operations Research: A Rich Source of Materials to Revitalize School Level Mathematics." In *Proceedings of the Fourth International Congress of Mathematical Education*, 1980 (to appear).

[20] Rene Thom. ""Modern" Mathematics: An Educational and Philosophic Error?" *American Scientist* 59 (1971) 695-699.

[21] Jean A. Dieudonne. "Should We Teach "Modern" Mathematics?" *American Scientist* 61 (1973) 16-19.

Teaching and Learning Mathematics

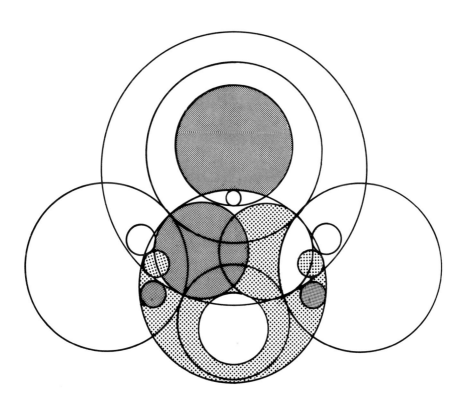

Avoiding Math Avoidance

Peter J. Hilton

The 1970's could well be described as the decade of anxiety. The action of the OPEC countries in quadrupling the price of oil in 1973 signalled the beginning of an era of profound economic uncertainty for the advanced industrial nations of the world; and the decade closed with 52 American citizens held hostage in the United States embassy in Teheran and more than 50,000 Soviet troops deployed in Afghanistan in support of a puppet regime. Politically and economically, the decade of anxiety could be the precursor of the decade of despair.

But it is not only in these domains that we find people anxiously surveying the circumstances of their lives. Advancing technology also brings its hazards. On the one hand, there is the ever-present fear of an irreversible contamination of the environment—from aerosol sprays destroying the ozone layer, from the leakage of dangerous nuclear waste and other accidents at nuclear power plants, from exhaust fumes polluting our atmosphere. On the other hand, advancing technology calls for better understanding of scientific processes and better ability to handle machines, and many of our citizens feel themselves utterly inadequate in the face of the challenge. Further, social and economic factors are combining to put more women than ever before into the work force, and many of these women are seriously handicapped by their inability to carry out the type of

Peter J. Hilton is the Louis D. Beaumont University Professor at Case Institute of Technology and a Fellow of the Battelle Seminars and Studies Program. He received his M.A. degree from Oxford University in 1947 and Ph.D. degrees from Oxford University in 1950 and from Cambridge University in 1952. He has taught at Cambridge, Manchester, Birmingham, Cornell, ETH (Zürich), NYU, and the University of Washington. Hilton is currently Secretary of the International Commission on Mathematical Instruction, and was Chairman of the United States National Research Council Committee on Applied Mathematics Training. His principal research interests are in algebraic topology, homological algebra, and mathematics education. He has published 15 books and over 200 articles in those areas, some jointly with colleagues.

quantitative thinking required by the job to which they aspire. Thus, in their daily lives, in their search for employment and their hopes of advancement, our citizens are beset by problems created by a combination of changing technology and defective education. If this is not to be the decade of despair for education, we must solve the problem of providing an education that meets the needs of today's—and tomorrow's—citizens. Obviously, the prime educational problem, viewed from the perspective of the issue of advancing technology, is that of mathematics education.

Unless one takes the pessimistic—and quite unjustifiable—view that the majority of human beings is incapable of mastering an elementary mathematical argument or applying an elementary mathematical skill, one is bound to conclude that there *are* serious defects in mathematics education. That the results are grossly unsatisfactory is attested by the prevalence, in our contemporary society, of programs to help adults overcome their aversion to mathematics. Sheila Tobias has formed an independent organization called *Overcoming Math Anxiety* and has written a book with that title [5]; likewise, Stanley Kogelman has formed an organization called *Mind over Math*, with similar objectives, and has also written a book (with Joseph Warren) with the title of his organization [4]. These are but two examples of the numerous initiatives being taken, up and down the country, to deal with a problem which is by now endemic and which is viewed as approaching crisis point.

All the experts agree that the source of the problem is not to be found exclusively in the nature of mathematics and of current mathematical education. There are strong societal forces militating today against effective mathematics education; some of these forces are particularly pernicious in their impact on female students. The analyses made by Elizabeth Fennema, Lucy Sells, Sheila Tobias, and others, of these forces are compelling—and alarming: we would add here that one of the strongest forces opposing good mathematics education today actually opposes *all* effective education. It is the search for instant satisfaction, for instant gratification. Education is a slow, gradual, cumulative process and its rewards are largely of the long-term variety. Today's young citizens are impatient and do not wish to work hard now for an enrichment in the future. We will argue in this article

---

One of the strongest forces opposing good mathematics education today actually opposes all effective education. It is the search for instant satisfaction, for instant gratification.

---

that it is not necessary to postpone to a distant future the rewards of hard study—an enlightened mathematical curriculum can capture the interest of the student at any level and provide delightful experiences—but it remains true that the benefits of a successful education will be felt by the adult throughout his (or her) life, just as the unhappy consequences of an unsuccessful education will also be felt.

The quest for instant satisfaction

Two other forces opposing effective education, which are not especially related either to the problem of women or to mathematical education in particular, derive from the quest for instant satisfaction. The first is the belief that the prime purpose of education is to guarantee a high material standard of living. Something seems to have been lost when the torch of universal enlightenment was conveyed from the old world to the new—the realization that, through education, our souls are enriched and our individual natures and attributes allowed to flourish. Of course, we expect, as a by-product of our education, to have available to us a variety of possible rewarding careers. But it is not true in our society that there is a simple relationship (expressed by a large positive correlation coefficient) between the intrinsic interest of a job and the salary it carries; and our education, qualifying us for interesting and challenging courses, may indeed disqualify us for positions carrying high salaries but little else of real merit. If there are those who believe that the "simple relationship" referred to is proper and right, let them campaign to achieve it; but let us not distort education away from its true purposes so that it becomes a means to material prosperity and ignores the deepest needs, and potentialities, of the individual.

The second malign force opposing education is rank commercialism. And in no respect are the pernicious effects of the ubiquity of commercialism on the quality of our lives, and the effectiveness of our education, more apparent than in our television programs. Not only is instant satisfaction "guaranteed" by meretricious advertisement; not only are glamour and success identified with superficiality and the tinsel achievements of paper heroes and heroines; but passivity becomes the natural mode of participation and the challenges of reality are replaced by the dubious charms of synthetic excitement. Children are, on the average, spending more time in front of the "box" than in the classroom; and the school is being asked to compete (at a staggering financial disadvantage!) with the soap operas, the endless series, and the rest in claiming the attention and the interest of the child. Here there are solid grounds for pessimism, indeed. Were the Soviet leaders as subtle as some think them to be, they would be giving clandestine support to commercial television companies rather than to subversive pseudomarxist movements.

In the face of these forces (and we have by no means listed them all—the inadequate funding of education, the inadequate remuneration of teachers, the lack of high public esteem for the teaching profession are some other forces doing their deadly work to cripple education), it is no wonder that there are such grave problems! But to them must be added serious faults in mathematics education itself. We will be devoting the rest of this article largely to a discussion of those more particular faults, but first we must make a point which we regard as of great importance in getting the right perspective on the problem.

The point is a simple one—indeed, it is obvious. With all the difficulties which education faces, coupled with the special problems of mathematics education, math avoidance is no pathology! Many of those trying to deal with the problem use the language of therapy—Sheila Tobias made "math anxiety" virtually a household word; the term "mathophobia" is also current; there are "math clinics" to help people overcome their anxiety; and the authors of *Mind over Math* give their names on the dust-cover of their book as "Dr. Stanley Kogelman and Dr. Joseph Warren," for all the world like two medical practitioners announcing a sensational new cure of a dreaded disease. This language is not only inappropriate, but gives a dangerously false impression, for it suggests that each individual exhibiting the behavior of math avoidance is suffering from some special affliction. The individual is liable to interpret the cause as being somehow to his or her discredit ("I was always very stupid about math," "Math was my worst subject," "Somehow I could never do math"), and thus to lose faith in his or her ability to achieve any reasonable level of competence. (It is also true that many people, successful in some sphere of life, may announce proudly that they are incompetent in mathematics. This inverted snobbery is itself a barrier to the spread of mathematical understanding.)

Bad mathematics—and good

In reality, so much that goes into the traditional mathematical curriculum, especially at the elementary level, is deplorable—it is not good mathematics, it is pointless and it is dull. To try to avoid it is, indeed, a thoroughly healthy reaction. At best, this material is suitable for a machine and it is a supreme irony that, dedicated as we are as a nation to the concept of the rights of human beings and the glory of their individualities, we devote a major part of our educational effort to turning our children into efficient machines (for obtaining accurate answers to mechanical problems), and to destroying their individuality by homogenizing and standardizing their responses. Even the most dedicated and skillful teachers—and there are very many such—must quail before so formidable and unnatural a task as that of rendering enjoyable the execution of the long division algorithm! But the problem is further compounded by the fact that many of our elementary teachers are *not* skillful in mathematics, and do not enjoy the subject themselves, and do not feel comfortable with it. There are so many subtle ways in which a teacher may communicate her attitudes to her students—ways of which she may not always herself be aware—and the effect on the child is obvious. The child (especially, perhaps, the female child), for whom the teacher provides a model to be emulated, becomes aware of the teacher's own attitudes—and adopts them. Distaste for mathematics is thus easily transmitted. It is an ironical reflection on this whole problem that children get turned off mathematics without ever having met mathematics—as the mathematician understands the term. What the child dislikes is rote calculation and the inculcation of arithmetical skills unrelated to the child's world. Thus if we are to try to improve the effectiveness of our teaching of mathematics, we must surely give emphasis to the naturalness and the usefulness of what the child is learning. Sadly, this tends not to be the emphasis in the elementary school—though there are, of course, many encouraging examples to show unmistakably what can be achieved with an enlightened curriculum, based on an understanding of a child's natural interests, curiosity and attitudes, and taught by a teacher who understands why it is important to learn how to approach a problem mathematically. But, in the majority of cases, as we have said, a subject misleadingly called "mathematics" is taught largely as a series of drill exercises; it thus rapidly becomes an unpopular subject, and the contrast between the treatment of this subject and of the others which the child learns may very well serve to confirm the habit of avoidance.

The "Back to Basics" movement . . . seems to be trying to do something which is not only not worth doing, but something which positively should not be done.

The mechanical approach to which we allude is best exemplified by the "Back to Basics" movement. This movement derived its momentum from the general dissatisfaction with the outcomes of the reforms initiated in the late 1950's and 1960's, reforms often characterized as the New Math. We cannot claim that the New Math was a success; but those who criticize it should keep various points in mind. First, the authentic New Math only reached a minority of our students (perhaps 15%), so that its introduction cannot be held responsible for the statistics of declining performance seized on by the advocates of "Back to Basics." Second, such evidence as we have tends to show that the New Math was more attractive to students than the Old Math. Third, what the New Math was *trying* to do was undoubtedly worth doing.

The "Back to Basics" movement, on the other hand, seems to be trying to do something which is not only not worth doing, but something which positively should not be done. What these enthusiasts of reaction want to take our children back to never was "basic" and is far less fundamental now than it ever used to be. With cheap hand-held calculators available it is manifestly absurd to be concentrating on those dreary mechanical calculations which bore children into permanent aversion. Of course, children should learn their "number facts;" but this is not to say that they should learn them as a pure effort of memory, unrelated to any task for which the number facts may be pertinent. There are attractive programs for the elementary grades in which the appropriate emphasis is given to the learning of these facts as a means of doing mathematics effectively; but, in these programs, there is a strong element of fun in the learning, and, typically, there are mathematical games to be played (an integral part of the program), in which the child plays the game more efficiently and joyfully if the number facts have been firmly grasped. (The author, from his own experience, would particularly commend in this respect the program *Real Math*, published by the Open Court Publishing Company; and the Comprehensive School Mathematics Program developed at CEMREL, St. Louis.)

The ills of mathematics education

Let us now be more specific in our diagnosis of the currently prevailing ills of mathematics education and our recommendations for possible remedies. We want to discuss both the education of the child and the education of the adult; but we must point out that these are distinct, though, of course, related, problems. First, then, let us look at the pattern of early mathematics education and see what features of that pattern are most conducive to math avoidance among those exposed to it.

Broadly speaking, these adverse features may be described as *dehumanization, artificiality, authoritarianism*, and *dishonesty*. Dehumanization is exemplified by rote calculation and memory dependence; artificiality by spurious applications and contrived problems; authoritarianism by rigid application of formal procedures; and dishonesty by a lack of candor, by unjustifiable attempts at motivation, by imprecise statements and, as with artificiality, by phony applications. We have given elsewhere [1] detailed instances of these defects; here we are content to refer to only one, the teaching of the addition of fractions. Sometimes this is justified (for example, in [3]) by a reference to a future topic which many of the students will never see; sometimes it is taught as a skill to be learned unquestioningly; sometimes "real-life" problem situations are presented which are alleged to warrant a mastery of the addition of fractions—but either the real-life problem would not, in fact, involve fractions or it would not be solved by any one facing it by using the model of fraction addition or subtraction. Thus we find in one text "John swims 22 2/3 meters on Monday and 23 4/7 meters on Tuesday. How far has he swum on the two days?" The application is spurious because John would never *know* that he had swum these strange fractional distances! He would know the distance he had swum in *decimals* of a meter, and the addition of decimals is easy. The pedagogy is dishonest, because the attempt is being made to convince the child that he or she needs to learn how to add fractions in order to be able to answer interesting questions—but the question, honestly presented, would never have involved fractions. In another text we find "You have 2/3 of a cup of flour and you need 3/4 of a cup. How much flour must you add?" In the unlikely event of the cook finding himself or herself in this situation, he or she would obviously simply fill the cup to the 3/4 mark —no self-respecting cook would subtract 2/3 from 3/4, measure out 1/12 of a cup, and add that to the 2/3 of a cup already measured out.

Tests tyrannize us—they tyrannize teachers and children. They loom so large that they distort the teaching curriculum and the teacher's natural style; they occur so frequently . . . that they appear . . . to be the very reason for learning mathematics.

Let these examples suffice. What should we do to improve the situation? The answer is simple and obvious: avoid these glaring blemishes in the standard pedagogy! But it is not so easy in practice. We have given many reasons for the inertia in the system, for the remarkable stability of those practices which militate against effective mathematics education. Let us be explicit about one further potent factor in the preservation of the status quo —the standardized tests. These tests, beloved of (some) educational psychologists and (many) educational administrators, superimpose a further degree of artificiality on that which is already present in the curriculum. They force students to answer artificial questions under artificial circumstances; they impose severe and artificial time constraints; they encourage the false view that mathematics can be separated out into tiny water-tight compartments; they teach the perverted doctrine that mathematical problems have a single right answer and that all other answers are equally wrong; they fail completely to take account of *mathematical process*, concentrating exclusively on the "answer." Particularly perverse and absurd is the multiple-choice format. I have been doing mathematics now as a professional for nearly 40 years and have never met a situation (outside finite group theory!) in which I was faced with a mathematical problem and knew that the answer was one of five possibilities. Moreover if faced, artificially, by such a situation, my approach would, and should, be quite different from that in which I simply had to solve the problem.

Tests tyrannize us—they tyrannize teachers *and* children. They loom so large that they distort the teaching curriculum and the teacher's natural style; they occur so frequently, and with such dire consequences, that they appear to the child (and, perhaps, to the teacher) to be the very *reason* for learning mathematics. (Indeed, I have seen the argument in print that long division must be learned because it appears on the tests!) Perhaps there are some good tests—for the rest, I say, down with them!

Positive principles of good teaching

It might appear so far that the principles of good teaching are largely negative—avoid the errors of the traditional course! There are, however, important positive principles worth enunciating. Let us list some: (a) there should be frequent reference to genuine applications of natural interest to the student—an application is genuine if it deals with a real-life problem and actually *requires* the mathematics under study for the solution; (b) the students should be encouraged to find mathematics in their own lives outside the classroom; (c) mathematics should be fun—this aspect can be developed through mathematical games and by exploiting the child's penchant for fantasy; (d) there should be an emphasis on problem-solving— but this should motivate the students toward mathematical understanding and skill development and not, of course, presented as an alternative to these; (e) the students should learn to think mathematically, and not merely to develop mechanical skills; (f) mathematics should be presented as a unity and not as a set of isolated disciplines (for example, the study of quadratic equations should include some algebra, some graphing, some geometry, and some estimation and approximation); (g) mathematics should be seen as labor-saving and not labor-making—a particularly good opportunity is furnished to make this point convincing by the availability and intelligent use of the hand-calculator; (h) the students should understand the roles of probability and statistics in determining intelligent behavior. To keep this article within reasonable length, I have not given examples of each of these principles. But the reader should easily be able to supply such examples—and counterexamples!

> *Unless the teaching is improved at the elementary level, the battle is really lost. We are never going to solve our problem by remedial strategies at the higher levels—we can only alleviate it.*

So far we have focused attention on the elementary grades. However, the principles apply with equal force at the secondary level (and almost equal force at the undergraduate level). Examples of bad pedagogy at the secondary and undergraduate levels are given in [1]; our reason for devoting most of our attention in this article to the elementary level is simply that, unless the teaching is improved at the elementary level, the battle is really lost. We are never going to solve our problem by remedial strategies at the higher levels—we can only alleviate it.

What then are the conclusions for the teaching of adults? We have principally in mind here those "math avoiders" forced back into the attempt to learn elementary mathematics by the exigencies of the job market or simply by the desire to cope better with the demands of modern life. Our community colleges are full of these people and will get fuller; our four-year colleges and universities are beginning to devote a major effort to remedial teaching, and will find themselves having to devote an increased effort. An increasingly large proportion of the students proceeding to higher education lack the necessary mathematical prerequisites and are enrolling in pre-calculus courses. The numbers are being further swollen by the new demand of many disciplines—conspicuously, the social sciences— for a mathematics qualification for their majors. Continuing education is going to be an increasingly prominent part of the service provided by post-secondary education and, within continuing education, mathematics courses are going to be in huge demand. What hope can we offer to solve this gigantic problem?

Let us speak only of the problem of teaching *mathematics* to adults, while recognizing that good counselling services are certainly going to be essential for many of our adult students. Then one principle stands out above all others— *the students must be treated as adults*. (Students must, of course, be treated with respect at all ages!) This means, obviously, that a rehash of their early experience of learning (or not learning) mathematics is going to be a dismal failure. It means that instruction based on the belief that adult students have ample, accurate and available memories, ready to store unassimilated and uncomprehended facts and mechanical procedures, will be even more disastrous than in the case of the teaching of children— for the memory capacities of adults are not nearly as good as those of children and are, moreover, much affected by the discrimination they have acquired in maturity. It means that dishonest motivation is going to be quite counter-productive.

Is there hope?

But there is hope! Enlightened, honest teaching, reflecting the positive principles we have already enumerated for early education, can overcome diffidence and lack of skill and understanding, and inculcate confidence and competence. Adults can be made to understand that they habitually execute mental tasks (like planning a vacation trip, budgeting their weekly expenditure, determining how much insurance to carry) far more subtle and sophisticated than those before which they tend to quail if recognized as involving mathematics. We must teach these people good, useful mathematics in a low-key, relaxed style, free from authoritarianism, so that they come to feel not threatened by it, not daunted by its unfamiliar symbols, but comfortable with it and convinced that they are now better able to reach rational decisions and cope with our complicated world. (These principles are, we hope, implemented in [2].) Mathematics will then become not a barrier, but the key to unlock the door to a fuller life.

Finally, we must say a word about the place of this essay in a symposium devoted to the future of mathematics. We have talked of the future (and past) of mathematics education—are there implications for the future of mathematics itself? Let us be content to give one reason why the answer is positive. If mathematics is to flourish it is essential that there be an adequate supply of academic positions for mathematicians. The supply of such positions depends on their being large enrollments of students. There will be—indeed, there are—large enrollments, but these students increasingly require fairly elementary courses. Mathematicians must therefore demonstrate an enthusiasm and a capacity for teaching such courses. This requires that they understand both the motivations of these students and the sources of their difficulties. The traditional training of a mathematician totally excluded such an awareness. Thus we are forced to conclude that, if mathematics, and the mathematical profession, are to remain healthy, there must be added to that traditional training (no substitutions allowed!) the component of real concern for, and skill in the teaching of, relatively elementary courses to students whose principal educational interests lie outside mathematics. Such a reorientation may not be the most formidable problem facing mathematics education today—but it is, as we mathematicians say, highly nontrivial!

References

[1] Peter Hilton. "Math Anxiety: Some Suggested Causes and Cures." *Two-Year College Mathematics Journal* (1980) 174-188; 246-251.
[2] ——— and Jean Pedersen. *Fear No More: An Adult Approach to Mathematics*, Vol. I. Addison-Wesley, 1981.
[3] Morris Kline. *Why the Professor Can't Teach*. St. Martin's Press, 1977.
[4] Stanley Kogelman and Joseph Warren. *Mind Over Math*. Dial Press, 1978.
[5] Sheila Tobias. *Overcoming Math Anxiety*. Norton, 1978.

Learning Mathematics

Anneli Lax and Giuliana Groat

A year ago we undertook the task of designing a mathematics course for freshmen who are admitted to a liberal arts college, but who are seriously limited by deficiencies in their pre-college mathematics education. We are not alone; many of our colleagues throughout the nation are involved in similar attempts and presumably are as naive as we are in matters related to this new assignment, such as the psychology of learning, education in elementary and secondary schools, and methods of evaluating students, teachers and courses.

We have, thanks to required courses in our college days, a dillettantish acquaintance with the ideas of Socrates, Rousseau, Montessori, Dewey and Piaget, but we have not embraced their particular theories of education or anyone else's. One of us has had the good fortune of observing Courant and Pólya at close quarters and has tried to analyze the secrets of their pedagogical success; we concluded that their methods like those of any extraordinary teacher defy analysis and cannot be bottled. During the last year we have read a great deal about mathematics instruction. We have found many positive contributions, but no really new ideas. A possible exception is the implications of recent brain research on understanding the

Anneli Lax is Professor of Mathematics at New York University. She came to New York from Europe at age 14 and attended the last two years of high school in New York City. She received her B.S. degree from Adelphi College in 1942 and began her part–time graduate studies at NYU in 1943 as a student of Richard Courant. Lax worked and studied with interruptions, receiving her M.A. in 1945 and her Ph.D. in 1955, both at NYU. After working off and on as a research associate and part–time instructor, she formally joined the faculty of NYU in 1961 and has been professor of mathematics since 1972. She has been editor of the New Mathematical Library ever since this series of monographs was created in 1958.

Giuliana Groat is Coordinator of the Basic Mathematics Program at New York University. She received her B.S. degree from Rollins College in 1968, and her M.S. from NYU in 1975. Since then she has been a part-time instructor in the Mathematics Department at NYU, administering and teaching in the department's Tutoring Program.

differences between ways adults and children learn and on the nature of thinking processes in general. Nevertheless, our recent thoughts and our considerable experience in teaching, tutoring and observing undergraduates have led us to form some strong opinions.

The purpose of this article is to share some of our views and findings with others who want to help ill-prepared high school graduates cope with their college studies or who want to improve pre-college training. We do not claim to address the entire problem here, nor do we recommend one particular approach. Effective programs cannot be transplanted from one university to another; student populations and resources vary, and each program seems closely tied to the personalities of the individuals who developed it.

Traditionally, university professors are expected to pursue research and train students for advanced work. They may enjoy introducing students to various fields, guiding them and sharing their ideas, but they generally do not consider it their responsibility to go back to square one in order to establish the necessary foundations, nor to give high pressure sales talks to interest indifferent students. This is not snobbishness; they simply do not think it their job, nor do they know how to go about it. If it suddenly became their job, some would probably do well, for they have at least one of the pre-requisites of a good teacher: deep, solid involvement with their discipline. But they could not simultaneously fulfill their primary function. Thus, while university teaching and research go hand in hand when

While university teaching and research go hand in hand when students are ready, they part company more and more as the gap widens between what students can do and what universities can offer.

students are ready, they part company more and more as the gap widens between what students can do and what universities can offer. The most efficient way to resolve the dilemma in the long run is to insist on adequate pre-college training; in the meantime the gap must be bridged by efforts within the university.

In facing the issue of why so many students entering college today have so little command of their language and of mathematics, we came upon a puzzling conflict. Certainly the broader and more solid a student's early training, the more options will remain open to him and the more versatile he will be in adapting to and shaping his world. However, our educational system seems to have become an instrument serving a society which requires certain diplomas and licenses for future employment. A college degree has become a non-negotiable requirement for beginning many careers, and schools participate in this sorting process. Today, when social, economic and cultural needs are in constant flux, when education should be providing a broad intellectual and methodological foundation, instruction is geared instead to short term objectives such as high scores on standardized multiple choice tests. This focus sacrifices involvement with the subject matter and does not leave room for sufficient attention to thought processes central to all studies.

Standardized objective multiple choice tests . . . neither set nor uphold standards.

Since devices such as SAT's and GRE's are used to rank students and influence educational decisions, it seems reasonable to ask if these tests can be relied on to identify students' strengths and weaknesses. We think not. Standardized objective multiple choice tests are primitive, much abused measuring devices. They neither set nor uphold standards; rather, they standardize (in the sense of mass-produce) the acquisition of skills. Nor are they objective in the sense of being independent of the subject's background. For example, in trying to rely as little as possible on language, some mathematics tests avoid word problems—an important indicator of mathematical ability—and substitute mathematical jargon comprehensible only to those who happen to have learned this particular jargon before taking the test. They may understand the mathematical concept without understanding the jargon; the test scores will never tell. Since intelligent students cannot be prevented from optimizing their test scores, they develop ingenious test-taking strategies, often by means of sophisticated mathematical thinking and estimating. We are nurturing a mathematical skill, but a rather narrow one.

Mathematics as a threat

Why does mathematics, among all the disciplines taught in schools, become so particularly threatening to so many students? Some of the reasons have to do with the nature of the subject, its ancient roots, its accelerating growth, and the claims for its all-pervasive applicability. Some have to do with mathematics teachers, with their own grasp of the subject matter, with the pressures exerted on them by a bureaucratic school system, and with their perception of their role in meeting society's needs for people with mathematical training. Last but not least, some have to do with attitudes students develop in school towards mathematics. Let us look at each of these in some detail. Mathematics is a way of thinking that requires considerable concentration. Its classical results have evolved over centuries and are presented as a closed hierarchy of logical sequences. All human traces—common sense, intuition, guessing, false starts—have been obliterated. Moreover, mastering its basic techniques with some facility, like learning to play a musical instrument, requires lots of practice with arithmetic operations and manipulation of symbols. Its exposition employs a rather special non-colloquial language: new words are defined, and known words have their meanings restricted or changed. This precise, denotative language is rarely reinforced, since the objects denoted are abstract and seldom connected with life outside the class.

The more anxious a teacher is, the more he retreats into the trodden path of memorized rules and rote instruction.

Most pre-college teachers find mathematics difficult. They are made uncomfortable by topics they don't thoroughly understand. Some feel threatened by students who ask questions or who seem to grasp and apply a new concept quickly or who explain a strategy different from the teacher's in solving a problem. The more anxious a teacher is, the more he retreats into the trodden path of memorized rules and rote instruction. Besides, he feels pressure to cover the material in the syllabus and can't take the time to listen to a student's often unclear explanation, let alone encourage the class to discover methods of solutions or reasons for the rules. After all, the teacher will be judged (by his superiors) by the performance of his students on tests. He is not aware that a more contemplative stance, a closer look at the ways different students think and learn, a class atmosphere where people dare to explain their ideas, might eventually lead to better student performance even on primitive tests. He observes that most students have trouble expressing themselves clearly and understanding concise statements, but he does not want to use his mathematics class for what he considers training in language. Even teachers originally committed to their discipline find it difficult, under the pressures of a bureaucratic school system, to stay in touch with that discipline and to teach it with some enthusiasm. Keeping up with current applications of their subject requires even more effort. Many teachers experience a growing feeling of inadequacy *vis à vis* mathematicians working in industry or at universities. These teachers gradually isolate themselves from their discipline. The topics they teach also become isolated not only from the rest of our world, but from each other. Courses are divided into algebra, geometry, trigonometry, calculus, etc., and so few connections are made that it is not uncommon for students to forget last term's material in this year's class. One of the most natural connections between topics, the mental processes brought to bear on each, is neglected because results rather than processes are emphasized. Neither the power nor the enjoyment of mathematics can be appreciated under these circumstances.

There are, of course, some excellent teachers and school administrators who do not conform to this bleak, simplistic sketch. We have seen remarkable mathematics teachers in action; we have encountered excellently trained, inquisitive students who stimulate our classes. There are active math clubs in many high schools; there are programs built around peer tutoring; there are science fairs and mathematics competitions that encourage collaborative creative efforts. Many of these flourish under the guidance of an inspiring teacher. Nevertheless there is great danger that such people will become rare as more and more products of narrow-goal training enter the profession.

Questions and answers

Student performance in mathematics can be quantified by the percentage of right answers to problems. It is easier in mathematics than in other subjects to define "right answer" unambiguously and, unfortunately, independently of the processes that lead to them. When instruction paves the path to right answers with rules to be memorized, students see no room for their creative contributions. They are not aware that mathematics has to do with thinking; they associate it with a body of formulas and algorithms to be memorized. They feel that mathematics is used more than other subjects to assess pupils, that those who are not good at manipulating symbols according to memorized rules are found wanting. Thus many students begin to develop grave doubts about their own abilities. The most vulnerable develop psychological blocks and anxieties, often exacerbated by teachers who suffer from similar anxieties.

While competition for a high place in the ranking is fostered in all subjects, in mathematics getting a quick right answer establishes a pecking order instantaneously. The right-answer-getters are "smart," the others are considered "dumb" and, in the eyes of many teachers, are candidates for remedial instruction. In order not to fall victim to the contempt of their peers, students hide their ignorance by not asking questions in class. (This behavior is not restricted to mathematics classes. We are told that it begins in first grade where some children enter able to read, others not; the latter feel inferior and learn to hide the very ignorance school is supposed to eliminate.)

In mathematics not asking questions has particularly dire consequences. Subsequent lessons on the same topic may depend logically on the very point somebody missed but was afraid to ask about; so he is doomed to not understanding a whole sequence of topics—a demoralizing experience. Moreover, the source of his confusion is less likely than in other subjects to be identified outside class because mathematics, as usually taught, has fewer connections with ordinary activities; nothing is likely to illuminate the murky point. Still another effect of not asking questions is that all members of the class are deprived of practice in articulating questions and answers and of collaborating with one another when solving problems. We cannot afford to neglect the art of clear articulation, for this is how others learn about our mathematical thought processes and how we organize them for our own purposes. (A fringe benefit of learning how to give a clear verbal description of a method is that making up clear instructions to program a computer will then pose no difficulty.) Again, the least secure teachers are the ones most likely to discourage dialogues instead of using them to enhance clear thinking and the effective use of language. Adequate language training would also address the most common complaint we hear from good as well as weak mathematics students: "I can't do word problems."

A consequence of mathematics instruction under pressure of time and with a tightly structured, isolated syllabus is that most students believe there is just one way of obtaining a result. If an adventurous spirit thinks of another, he or she suspects it must be wrong. Few students are sufficiently secure to insist that they be heard, or sufficiently articulate to argue persuasively for their method of solution. So the "this is how you do it" approach prevails and gives mathematics its reputation for being dull, dead, and practiced by non-creative people.

Movements and reactions

There have been a number of promising developments in the last few decades, in curriculum reform, in identification and treatment of psychological blocks, in artificial intelligence and instructional technology, and especially in computer aided instruction.

The most drastic curriculum reform in the last thirty years is generally known as "the new math." Its main efforts consisted of emphasizing some of the structure underlying arithmetic in the hope that an understanding of some fundamental principles and their logical consequences would replace dull routine drill with deeper insights which, in turn, would lead to wide generalizations and intellectual challenges. One of the most beneficial by-products of the new math consisted in bringing a large number of teachers back into mathematics, at least temporarily. It did its greatest harm in school systems that imposed it on teachers who lacked a solid knowledge of mathematics. It also did some harm in the hands of zealous well-prepared teachers who failed to take into account that young children often have not yet had the concrete experiences necessary for modelling abstractions, and that children see no need for proving things that seem obvious to them. The new effort was often couched in a pedantic, stilted language that removed mathematics even further from connections with the rest of the world. No concrete applications were offered to bridge the schism between abstraction and reality. In many instances, teaching of traditional rote methods was replaced by teaching new rote methods in a new vocabulary. Many game-like activities formerly used to drill children in arithmetic were abandoned, with the result that masses of pupils grew up without computational facility. In many instances, instruction in the new math was not made sufficiently flexible to accommodate pupils who like first to learn the rules of the game, and who only later ask the why's and wherefore's, discover patterns in their games, and are ready to think about them.

The recent "back to basics" movement has been directed mainly at cutting out some frills and nonsense, and getting schools to concentrate on their main business. Its ranks include many critics of the new math. The part of the movement which insists that schools provide a solid broad common base so that options remain open for later specialization deserves support. But its fringe elements who insist on returning to a particular set of old curricula and teaching methods would keep elementary education divorced from the society it must serve.

A vigorous interaction of analytic and heuristic thinking seems to us most fruitful, especially if people learn when to apply which.

We believe that it is important to expose students to both, to a discipline with a rigorous structure learned in a logical sequence, and to activities involving much information gathering with flexible frames of reference continually modified as more information is processed. A vigorous interaction of analytic and heuristic thinking seems to us most fruitful, especially if people learn when to apply which. A mathematics syllabus consisting of a narrow vertical structure poses great risks to the student who misses a step near the bottom. Each subsequent piece is incomprehensible to him.

There seems to be pretty general agreement on roughly which mathematical topics should be treated in elementary schools. Educators know from experience the optimum time in a child's development for learning certain things. However, they do not always take advantage of this knowledge. For example, it is well-known that young children can learn to speak a foreign language without an accent (a talent they lose in adolescence), and with practice, will continue to speak it so; yet languages are rarely taught at the optimum time. Arithmetic generally is taught at roughly the proper time, but if elementary computational skills do not become part of a child's resources, and if he must learn such skills later, he does it quite differently, with much greater effort, and forgets them more easily.

Opinions on which topics should be taught in high school and in what order differ a great deal. There are guidance counsellors who steer ninth graders, especially girls, away from algebra into "general mathematics;" victims of this advice are later crippled, in college or out, by their inability to understand and use symbols. For students in college preparatory tracks, the standard fare is algebra, plane geometry, more algebra and trigonometry, usually served as separate dishes. Many schools offer calculus, some also selected topics from probability, matrix algebra, Boolean algebra and set theory to interested students. Analytic-, solid-, projective-, spherical-, non-Euclidean geometries have been almost entirely abandoned, and the old-fashioned synthetic treatment of Euclidean geometry in tenth grade has been greatly curtailed. The high school calculus course in the College Board's Advanced Placement Program is a great time and money saver for science-oriented students. It is also a stimulus for the mathematics teachers in schools that offer it. At the same time, it is an example of a vertical path along which students are pushed to give them a head start on the competitive road toward graduate or professional schools. Opponents of high school calculus courses object on two grounds: One is that many students exposed to calculus in high school know standard techniques, e.g., differentiating elementary functions, but fail to understand underlying concepts. As a result they often have delusions of competence and are bored in the early parts of college analysis courses. When the study of analysis makes greater technical and intellectual demands, many of these students fall behind. The other objection—especially if calculus is the only elective mathematics course in high school—is that it displaces such beautiful and important topics as complex numbers, conic sections, polynomials, number theory, combinatorics. All these can be introduced as extensions of topics in the standard syllabi of the first one and a half years, broadening a student's scope and providing a synthesis of mathematical ideas and methods he or she might never glimpse otherwise.

Anxiety therapy

Psychological blocks that prevent people from learning mathematics have recently received much attention from psychological counsellors collaborating with mathematicians in combating "math anxiety" and "math phobia." Skillful psychologists are able to identify and trace the origins of such blocks in students seeking help. The anxiety-causing trauma usually occurs early in life, and devastates the victim's self-confidence. Consciously or subconsciously, she or he develops self-protective strategies not conducive to learning mathematics. By the time such a student graduates from high school, negative feelings are deeply ingrained. If early trauma could be prevented, colleges might not need to make the large scale salvage attempts we are now witnessing, and the non-college bound population might enjoy greater mathematical literacy.

Perhaps the most effective salvage work is being done by people interested in bringing women and other victims of stereotyping into mathematics-related fields. It is particularly helpful to introduce mathematical topics (e.g., graph theory or combinatorics) that students do not associate with the sort of mathematics they used to hate. There are students, however, who feel successful only if they learn to master precisely those techniques that scared them most, and for these, a different initial approach is more helpful. In young children it may be their state of physical development that temporarily limits certain kinds of mental activities; in math-anxious adults, such limitations usually have psychological causes. In all cases it would help to first use accessible mental channels and then strive to open as many more as possible.

All math-anxious students seem to respond favorably to empathetic instructors who believe that their students can learn a certain amount of mathematics. Some mathematically sophisticated instructors have the required faith in their students but are unable to fathom the kinds of difficulties some novices have with the most basic elements in mathematical reasoning, for example, with matters of notation. Unless peers or tutors intercede, such teachers may begin to doubt their ability to get something across or they may lose faith in their students, or both. This is not a rare phenomenon and has marred the success of many an elementary college mathematics course.

Some teachers mistake people's run-of-the-mill anxieties for math anxiety. We have seen questions on math anxiety rating scales which many competent mathematics students would answer so as to be rated very math anxious. Yet these people are more at ease with mathematics than with any other subject. As their involvement with mathematics deepens and their performance improves, their other anxieties (e.g., those due to time pressures or to competition) may lessen. The danger we see is that some well-intentioned teachers may get so hung up on math anxiety that they neglect the seductive aspects of mathematics which lead students into the subject and away from their self-consciousness. Teachers with insufficient mathematical knowledge who themselves suffer from math anxiety especially tend to project their difficulties and substitute anxiety therapy for mathematics instruction. If this became widespread, it might lead to an unfortunate math anxiety backlash, counteracting many recent positive efforts.

Brain research

Those of us interested in how we, our colleagues, and our students think when confronted with a mathematical problem are fascinated with recent work on brain functions. We suspect that the implications of current brain research will support methods of teaching mathematics to which perceptive instructors have been led intuitively. For example, many topics can be introduced analytically, then illustrated geometrically, then elaborated on via other models related to the experience of some students. Those who understood the first presentation can see it from a different perspective without getting bored, and those who did not understand the first are given other opportunities to grasp the material. Additional connections are made when the new topic is applied to diverse situations, enlarging a student's store of potentially useful tools for solving problems.

An article on page one of the March 30, 1980 issue of the *New York Times* entitled "New Teaching Method Raises Hopes in Inner Cities" describes "Mastery Learning" as a "new teaching method that rejects the assumption that failure is an inevitable part of schooling." The article then reports on the success of a teaching technique (in any subject) consisting of identifying those students who did not understand the initial presentation of a topic and giving them a second chance by "shooting a different arrow" the next time around. Students who had grasped the first explanation were termed "fast learners." They are given "enrichment material" while that second arrow is being shot at the "slow learners." One teacher reported that he had some of the "fast learners" help their slower peers, and that they communicated more effectively than he could. Under "mastery learning," it was reported, "students achieve better, like school better, and attendance improves."

We were astonished that the technique of more than one way of skinning a cat was called "new." We wondered why only two ways were used, and why the people who learned from the first and second approaches were called "fast" and "slow" respectively just because of an arbitrary ordering of two different approaches. We fail to see why "inner cities" were singled out as target for "mastery learning." In mathematics classes, "mastery learning" requires a teacher who knows more than one way of presenting a topic; we hope this will not stall the project. Our own experience confirms that peer teaching is of enormous value to both the explainer and the listener. We wonder why it is used so sparingly.

The author of another recent article in an education journal discussed implications of recent brain research on arithmetic teaching. Here the thesis was presented that children with unusually great difficulties in arithmetic seem to suffer from a delayed development of their dominant hemisphere and should therefore receive separate remedial instruction, tailored to their needs. Again, the author is ranking different learning styles and would segregate children, thus depriving them of becoming familiar with somebody else's way of thinking and of the stimulating interaction of these diverse processes.

Teachers and technology

Computer-aided instruction has a number of advantages and disadvantages. When well programmed, a computer can relieve teachers of giving and correcting drill exercises, even of teaching many manipulative techniques. Students who react negatively to a teacher can turn to the computer; it instructs without emotional involvement and gives instant, impartial feedback. It shows no impatience no matter how often a student repeats a mistake. The computer's greatest asset is that it forces the student to become an active participant.

All visual aids and models—kindergarten blocks and bars, the number line, mathematical video tapes and TV screens—are extremely valuable, but it seems to us that students learn even more if they have to create their own images and models. Their ability to do this depends on their experience. Perhaps something like Papert's turtle can be used to do both, to furnish experience and encourage modelmaking.

There have been enthusiastic reports of successful use of computers in elementary schools, mainly in California. Again, the active involvement of the child and his ability to pick up the skill of communicating with the computer quickly (faster than most adults) seem to be the essential assets. Clearly, those who write ingenious programs deserve credit. In order to make this kind of instruction accessible to elementary school pupils on a large scale, one needs the necessary funds, facilities, and a good staff to get the most out of human-computer interaction.

In colleges, computers are being used in some remedial mathematics programs, especially in the "proceed at your own pace" variety. They are also used in regular mathematics courses, mainly in elementary calculus. In the hands of competent staff who know how to enhance student thinking by using computers, this works very well (see, e.g., the Ohio State University course).

Some educational psychologists hope to enlist the aid of microcomputers in determining a student's learning style and in diagnosing learning difficulties. This, together with research on brain functions and artificial intelligence, may lead to improvements in teaching and learning methods by giving objective support to our hunches about who learns in which way, and what sort of mental exercise will establish new ways of learning. At this stage, it seems to us, the judgment of good, observant teachers is still ahead of the inferences drawn from research on brain development, brain function and computer aided instruction. But we have no doubt that recent advances in research will have a deep impact on teaching, and we hope that researchers and teachers will talk to each other a lot.

Clearly much energy is focused on students with deficient mathematical backgrounds, on the reasons for those deficiencies, and on methods to remedy them. We have found that many efforts that are under way to attract people to mathematics base their pitch on the importance of the subject for careers and consumer decisions but neglect to point out what we consider a more compelling reason for studying the subject. Mathematics is

The acquisition and exercise of these mental powers [of mathematics] can be gratifying, engaging, and in some cases almost addictive.

a way of thinking which enables students to see unifying patterns in diverse contexts and to analyze complex situations. The acquisition and exercise of these mental powers can be gratifying, engaging, and in some cases almost addictive. If we can show this to students by doing mathematics with them, if they can gain some inkling of the intrinsic value of such an involvement, then they will have learned an extremely important lesson: that they, by using their very human faculties of thinking and caring, can gain some understanding of how things work and can often make them work better; that education is not only a means for survival and success in a ready-made world but that it is a gradual and continuing transformation—an end in itself—leading students to become adults who learn throughout their lives and who are capable of making their personal and very human mark on whatever situations they encounter.

Our look into probable causes and possible prevention of student illiteracy has led us from our own students to high school graduates in general and leaves us more convinced than ever that literacy and numeracy are essential also for the non-college bound. They too are entitled not only to the financial benefits of a good schooling but also to the excitement that education should provide.

In conclusion, we return to the starting point of our inquiry, the problem of preparing students for post-secondary mathematics education. The most difficult part of this task, we conclude, is to strike the proper balance between apparently antithetical concerns: to help students cope with their fears and improve their self-images, yet get them involved in doing mathematics, not just in introspection and therapy; to find out how each student learns best and give individual attention, yet prepare him to cope with professors and authors of texts who instruct in other modes; to have students acquire and enjoy facility with mechanical techniques, yet resist "programming" students to perform; to insist on clear thinking, and precise verbal arguments, yet not get stuck in lifeless pedantry; to exhibit the structure of our discipline, yet establish as many connections as possible with concrete experiences; to develop individual talents and strengths, yet keep in mind societal and career demands and expectations from our students.

Teaching Mathematics

Abe Shenitzer

Each person has at any given time a private reality, a private realm of the meaningful. To teach someone is to extend this reality, to enlarge it.

This broadening process involves the development of critical abilities and of an intense regard for ideas. The educated person will weigh, consider, evaluate, conclude and doubt. His crucial skill is his ability to acquire, to sift, and to use information with a view to reaching the deepest insight, the fullest understanding. Somewhat paradoxically, the educated person can manage his own education; he is autonomous.

Teaching cannot succeed unless, at each stage, we build on some prepared base, on some "lower reality," but a reality nonetheless. If a student is aware of the ubiquity of groups, then he may participate in a certain amount of study of group theory. If he has a grasp of elementary physics, then he may take some interest in differential equations. If he is impressed with infinite series, then he may also be impressed with efforts to improve their convergence. If we can get him to learn enough mathematics to understand the issues involved in the Bernoulli-d'Alembert-Euler-Lagrange debate which culminated in the Fourier Theorem, then the student may agree that mathematics is not a difficult, useless and meaningless subject, but a potentially exciting intellectual adventure.

Mathematics is a crucial component of our culture. It can and should make a signal educational contribution. If it fails to do so it is because, like all other subjects, it is all too easily subverted. The dangers are many: Its great ideas may be smothered by a mass of theorems.

Abe Shenitzer is Professor of Mathematics at York University. Born in Warsaw, Poland, Shenitzer graduated from Brooklyn College in 1950 and received his Ph.D. from New York University in 1954. He has been a staff member at Bell Labs, and has taught at Rutgers and Adelphi University. He has translated nine mathematics books from Russian and German into English.

> ... [Galois] searched textbooks and papers in vain for the great leading ideas. He was convinced that the failure to emphasize these ideas allows the intellectual core of the subject to be smothered by a mass of theorems. [1, p. 73]

It may suffer from a preoccupation with triviality,

> We believe that arithmetic as it has been taught in grade schools until quite recently has such a meagre intellectual content that the oft-noted reaction against the subject is not an unfortunate rebellion against a difficult subject, but a perfectly proper response to a preoccupation with triviality. [2]

from fragmentation,

> ... there is something wrong with the present system—something which is apparently being aggravated by the recent upsurge of mathematical research. I refer to complaints from students that the various courses we are offering them are too self-centered, and that their teachers were making no effort to interrelate these courses. And they were asking, how do all these specialities relate to one another, and what significance do they individually have for the bulk of mathematics? Where is it all going, anyway? [3, p. 481]

or from a distortion of educational objectives which produces "specialists, technicians without mitigating philosophic or reflective resources":

> ... among some of the most capable, research-wise of the new Ph.D.'s, can often be found the greatest lack of knowledge concerning the background and significance of their work, as well as abysmal ignorance of the reasons for doing it and of the general nature of mathematics. In short, they are uneducated specialists. If you ask them why they are specialists, the best reason they can give is that this is the way to get results which merit publication and hence a good job. [3, p. 482]

If the educational potential of mathematics is to be realized, then the attention that has to be paid to its technical aspects must be balanced by attention paid to its structural, historical, genetic and philosophical aspects.

We are by now fortunate in having a few textbooks whose approach to various mathematical topics is genetic or historical (e.g., [4], [5], [6]). My own comments in this essay illustrate the structural approach to mathematical topics. They are modest echoes of comments made by some of my teachers. (One of my teachers, for instance, began a course in geometry with the remark "Unlike Euclid, Hilbert realized that it is possible to study form without substance." Another, in a course in complex variables, remarked: "How does one find interesting classes of functions? We agree that analytic functions are important. But then so are harmonic functions. But harmonic functions are continuous solutions of the Laplace equation. This suggests that differential equations may be one source of interesting classes of functions.") Some of my comments take the form of examples illustrating a concept. Some deal with a concept or an idea which unifies seemingly unrelated topics. Some represent an attempt to isolate a key idea of a theory. Some are didactic in nature. The technical material involved ranges from the addition of fractions to Fourier's Theorem. This variety reflects my conviction that structural comments need not be restricted to a particular field of mathematics or to a particular level of exposition.

If the educational potential of mathematics is to be realized, then the attention that has to be paid to its technical aspects must be balanced by attention paid to its structural, historical, genetic and philosophical aspects.

I do not claim any specific significance for these illustrations. They certainly can, and frequently should, be replaced by a variety of other examples. But they should not be dismissed with the remark that they are obvious. Maybe they are for the teacher. But I know that they are not obvious for the majority of students.

Functions and equations

Bafflement may start one on the road to insight. With this in mind I sometimes ask my students the following question: Functions are presumably among the most important objects of mathematical study. If a function is given by a rule such as $y = 2x$ or $y = x^2$, then we know what output the function associates to any given input. But then, what is there to study?

All too often, the audience is stunned by the question. I point out that it is hardly possible to study that which is not somehow given. Once the shock of this discovery has worn off, we agree—after some Socratic dialogue—that if f is a function with domain of definition D, then the two questions we usually deal with first are:

(1) What is $f(D)$, the range of f?
(2) If a is in $f(D)$, what is $f^{-1}(a)$, the set of elements of D which f takes into a.

To solve the equation $f(x) = a$ is to determine $f^{-1}(a)$. If f is a polynomial then to find the roots of if is to determine $f^{-1}(0)$. The unifying power of this remark is obvious.

Questions: Can you think of functions for which these questions are easy to answer or difficult to answer? (Hint: consider Fermat's Last Theorem and the Riemann Hypothesis.) What can you say about the sets $f^{-1}(a)$ if f is a homomorphism? How can they be structurally related to $f(D)$ if the latter is a group, say?

"Pure" vs. "applied"

Instead of adding fractions according to the rule

$$\frac{a}{b} + \frac{c}{d} = \frac{ad + bc}{bd}$$

some students write

$$\frac{a}{b} + \frac{c}{d} = \frac{a + c}{b + d}.$$

The first rule is "right" and the second is "wrong." The authoritarian "it's the wrong answer" subverts a potentially rational activity and replaces the necessary liberating discussion with an obedience response. What is important is that the student must realize that as long as he is manipulating

---

The deep insight is that rules of operation . . . are, intrinsically, neither "right" nor "wrong." What may make them "right" or "wrong" is the uses to which we put them.

---

abstract symbols any rule is as good as any other rule. What makes one rule "right" and another "wrong" is the applications we have in mind. Thus if you are buying a/b pounds of sugar and then an additional c/d pounds of sugar, then the number of pounds of sugar bought is

$$\frac{a}{b} + \frac{c}{d} = \frac{ad + bc}{bd}.$$

On the other hand, if in one season a team played b games and won a of them and in another season it played d games and won c of them, then, in all, it played $b + d$ games and won a + c of them; in symbols,

$$\frac{a}{b} + \frac{c}{d} = \frac{a + c}{b + d}.$$

The same rule holds if we take a/b to mean a pair of numbers where a is the real and b is the imaginary part of a complex number.

The deep insight is that rules of operation may produce a more or less interesting system but are, intrinsically, neither "right" nor "wrong." What may make them "right" or "wrong" is the uses to which we put them.

Question: What reasons can you offer in support of the "rule of signs" for the multiplication of integers?

While on the subject of addition of fractions we might add that the issue is usually confused by consideration of greatest common divisors and least common multiples. Note that the rule

$$\frac{a}{b} + \frac{c}{d} = \frac{ad + bc}{bd}$$

says nothing of greatest common divisors.

Greatest common divisors come up in a natural way if we consider, say, the problem of finding the integer values of a for which the equation $14x + 21y = a$ has integer solutions x, y. The values in question are $a = 7k$, where k is an integer and 7 is, of course, the greatest common divisor of 14 and 21.

Question: Formulate the problem of solving $14x + 21y = 42$ in integers in terms of finding $f^{-1}(42)$ for a suitable f and solve it by first determining $f^{-1}(0)$. Does this suggestion remind you of what is done in linear algebra in solving a system of linear equations?

Mathematical atoms

Chemical analysis has its analogues in mathematics.

Consider the integers. The Fundamental Theorem of Arithmetic asserts that every integer not 0, 1 or -1 can be written in an essentially unique way as a product of primes. Thus the primes are the multiplicative building blocks of the integers. (The uniqueness part of the statement is one of the most frequently utilized properties of the integers.)

Consider the complex polynomials. The Fundamental Theorem of Algebra asserts that every complex polynomial is an essentially unique product of linear polynomials. Thus the linear polynomials are the multiplicative building blocks of the complex polynomials.

The Basis Theorem for finite commutative groups is another instance of a factorization theorem. It asserts that every finite commutative group is isomorphic to, that is, structurally identical with, a direct sum of cyclic groups each with a prime-power number of elements. Thus the cyclic groups of prime power order are the additive building blocks of finite commutative groups.

The Jordan-Hölder theorem exhibits the simple groups as building blocks for finite groups. Here the building process is the extension of a group by a group. And the complexity of the structure of finite groups can easily compete with that of the molecules in chemical compounds. (Of course, to make this analogy effective, one must hope that the students know what a chemical compound is. Unfortunately, such is not always the case.)

The formation of infinite sums of simple functions is a favorite function-building tool. At first sight it is inconceivable that there could exist simple additive building blocks for the class of all functions on an interval. Nevertheless, in 1822 Fourier announced his famous result to the effect that every function on an interval is the infinite sum of definite multiples of the simple functions 1, $\sin x$, $\cos x$, $\sin 2x$, $\cos 2x$, $\sin 3x$, $\cos 3x$, ... These "functional building blocks" have the great virtue of representing simple harmonic motions (exemplified by the motion of the bob of a swinging pendulum). The function-building role of the sine and cosine functions combined with their fundamental physical significance justifies the inordinate amount of attention lavished on them. (Actually, Fourier's claim was not quite correct. A modern result, with a sweep comparable to that of Fourier's, states that the Fourier series of a function f in $L_p(-\pi, \pi)$, $1 < p < \infty$, converges pointwise to f almost everywhere.)

Questions: Discuss the issues of uniqueness in the above examples. What are the multiplicative building blocks of the set of even integers?

Clarifying the deductive method

The deductive method is usually introduced in connection with the study of geometry. Since the early theorems of Euclidean geometry are "intuitively obvious," students are often baffled by the need to prove them, and may end up respecting neither the deductive method nor the truths obtained by its application.

The failure to distinguish between the formal side of mathematics and its sources of inspiration is an affliction that is not restricted to students of elementary geometry. Many students of advanced calculus cannot answer the question: Why prove the Intermediate Value Theorem? Isn't it obviously true? Again, many students of hyperbolic geometry cannot answer the question (asked in connection with the study of the Poincaré model of the hyperbolic plane): Isn't it absurd to call a circular arc a straight line? And is it surprising that the consequences defy common sense?

The resolution of these and of similar "paradoxes" is far more important than endless proofmaking, for it may save the students from acquiring the all-too-widespread conviction that mathematics is an elaborate pursuit of the meaningless and, at the same time, increase their appreciation of the deductive method.

> ... it is a sad reflection on the intellectual level of mathematical education that ... the mathematics student may get his degree without having heard about Gödel's ... monumental discovery ... widely regarded as one of the greatest intellectual accomplishments of the 20th century.

The study of "neutral" sets of axioms may be another way of increasing student appreciation of the deductive method. Here is an example of what I have in mind.

The group S_n of all shuffles (permutations) of a set of n objects is a "natural" object. So are the integers under addition, the rationals under addition, and the nonzero rationals under multiplication. Starting with these examples we might perhaps be able to "sell" the "neutral" group axioms to our students at an early time and to study the implications of these axioms to the point where we can prove, say, Lagrange's theorem—an entirely unanticipated result.

Speaking of the deductive method, it is a sad reflection on the intellectual level of mathematical education that, unless he takes courses in logic, the mathematics student may get his degree without having heard about Gödel or about his monumental discovery of the intrinsic limitations of the deductive method, a discovery widely regarded as one of the greatest intellectual accomplishments of the 20th century.

Mathematical sameness

If we sort a batch of nails by length then the nails which end up in the same box are "the same" from the point of view of length. If we sort them by length and thickness, then the nails that end up in the same box are "the same" from the point of view of these two qualities, and so on.

The mathematical sameness relations are called equivalence relations. What is striking about many important mathematical equivalence relations is that they involve groups in one way or another. Thus the equivalence of (positive) fractions is an instance of the equivalence of words in a free abelian group—that of the positive fractions under multiplication. Again, to say that two integers are congruent modulo n is to say that they belong to the same coset with respect to the additive subgroup of multiples of n. (A similar example is that of equivalence of Cauchy sequences of rational numbers in Cantor's construction of the reals.) Finally, an important class of equivalence relations involves groups in the very definition. One example of such an equivalence relation is similarity of matrices. Another example is furnished by the classification of, say, quadratic curves with respect to some group of transformations of the plane.

There is yet another, more vague, kind of sameness: Two problems may seem different but may involve the same central concept. For example, in each of the equations

(a) $ax \equiv b \pmod{n}$,

(b) $x^m \equiv a \pmod{n}$,

(c) $ax + by = c$, for a, b, c integers,

(d) $ax + by = c$, for a, b, c real numbers

the unifying concept is that of a homomorphism. In each of these equations we are looking for a coset of the kernel of the homomorphism.

Consider a different example. Polynomial equations $P(x) = 0$, x an unknown number, are studied even in very elementary courses. If we replace x with the differential operator D, we are led to the equation

$P(D)y = 0$. Here we are looking for functions $y = y(x)$ satisfying a linear differential equation with constant coefficients. Such equations are involved in many physical applications. The polynomial equation is certainly different from the differential equation. Nevertheless, if we know the solutions to the polynomial equation, then we can write down the solutions of the corresponding differential equation.

These two examples resemble musical variations on a theme or variants of a theme in comparative literature. Instead of relegating variants of a mathematical problem to different courses (number theory, linear algebra, etc.), we should make a point of considering them as different facets of the same problem.

Question: Can you think of other mathematical "variations on a theme?"

The "right" setting

The right setting for the study of factorization is an integral domain which is not a field; the study of factorization in a field reduces to a triviality.

The right mathematical setting for quantum mechanics is Hilbert space.

You cannot "do" analysis over the rationals. This is largely due to the fact that the rationals lack the property of completeness, that is, there are Cauchy sequences of rational numbers without a rational limit. The field of real numbers is complete, and so we have "real analysis." The field of complex numbers is complete and so we have "complex analysis." Since the field of complex numbers is not only complete but also algebraically complete (i.e., a complex polynomial of degree n has exactly n roots), it frequently offers special advantages. In the words of J. Hadamard, "the shortest path between two truths in the real domain passes through the complex domain."

The study of continuity in the setting of the calculus is cluttered up with irrelevancies which vanish when we shift to the setting of metric spaces (Fréchet, 1904) in which all that matters is distance.

In spite of their apparent generality, metric spaces cannot accommodate as simple a phenomenon as pointwise convergence of a sequence of functions in the sense that pointwise convergence cannot be induced by a metric. (There is no metric on the space A of functions on an interval whose convergent sequences are precisely the pointwise convergent sequences.) On the other hand, pointwise convergence can be induced by a suitable topology. (There is a topology on A which induces as its convergent sequences precisely the pointwise convergent sequences.) This fact alone constitutes an argument in favor of topological spaces and prepares us for the insight that it is topological spaces rather than metric spaces that provide the "right" setting for the study of continuity.

Questions: Why would mathematical physics "cave in" if it had to work with the rationals rather than with the reals or with the complex numbers. (Hint: Look at the definitions of such concepts as density, velocity, acceleration, etc.)

Can you think of situations in which the complex numbers are, in some sense, "better" than the reals? (Suggestions: Use complex numbers to show that the composition of two rotations, about the same center or about different centers, is again a rotation and find its center. Compare a^b in R and in C. Compare power series in R and in C. Read pp. 234-243 in S. Bochner's *The Role of Mathematics in the Rise of Physics* [7].)

Mathematical creation

The making of new mathematical systems out of given ones is a common mathematical activity. The equivalence classes of integers modulo n are combined to form cyclic groups and certain finite rings. Construction of the number system involves the chain of extensions $Z^+ \rightarrow Z \rightarrow Q \rightarrow R \rightarrow C$. Homomorphisms induce factor structures. Direct sums are a natural technique for linking two or more given structures, and so on.

... this conscious act of creation of mathematical systems is less than 200 *years old.*

What should be stressed in this connection is that this conscious act of creation of mathematical systems is less than 200 years old, and that the discovery of the existence of incommensurable magnitudes by the Pythagoreans some 2500 years ago led them not to the enlargement of what was effectively the system of rationals but to the downgrading of numbers as unequal to the tasks of geometry. The moral here is that what is often most revolutionary is the change of viewpoint.

In conclusion, let me say this. My remarks are intended as samples of a particular form of critical appreciation of mathematics. Critical appreciation must be a fixed component of the study and of the teaching of mathematics. Without it the student remains illiterate and the teacher is a mere purveyor of deductive tapeworms. Critical appreciation lifts mathematics from the level of a crossword puzzle to that of a component of culture. The uneducated specialist may be effective in a variety of ways but as a teacher he is bound to fail in the fundamental task which is (to quote Peter Hilton) to "strive to awaken in as many people as possible, irrespective of their chosen vocation, an awareness of the nature of our science, and its significance for our civilization, material and spiritual."

References

[1] H. Wussing. *Die Genesis des abstrakten Gruppenbegriffes*, p. 73.
[2] *The Report of the Cambridge Conference on School Mathematics*, 1963.
[3] Raymond L. Wilder. "History in the Mathematics Curriculum." *American Mathematical Monthly* 79 (1972) 481.
[4] Otto Toeplitz. *The Calculus: A Genetic Approach*. Univ. of Chicago Pr., 1963.
[5] Harold M. Edwards, *Fermat's Last Theorem. A Genetic Introduction to Algebraic Number Theory*. Springer-Verlag, 1977.
[6] C.H. Edwards, Jr. *The Historical Development of the Calculus*. Springer-Verlag, 1979.
[7] Salomon Bochner, *The Role of Mathematics in the Rise of Science*, Princeton U. Press, 1966.

Read the Masters!

Harold M. Edwards

> It appears to me that if one wants to make progress in mathematics one should study the masters and not the pupils.
>
> — N.H. Abel (1802-1829), quoted from an unpublished source by O. Ore in *Niels Henrik Abel, Mathematician Extraordinary*, p. 138.

It is as good an idea to read the masters now as it was in Abel's time. The best mathematicians know this and do it all the time. Unfortunately, students of mathematics normally spend their early years using textbooks (which may be, but usually aren't, written by masters) and taking lecture courses which are self-contained and make little or no reference to the primary literature of the subject. The students are left to discover on their own the wisdom of Abel's advice. In this they are being cheated.

The phrase "read the masters" can be interpreted in two ways. There are mathematical specialties today which scarcely existed 50 years ago, and in these specialties "reading the masters" would mean reading the works of the most important contributors. This is indeed a very worthwhile activity; it is pursued by the best workers in such fields, but even here students are often not directed to the primary literature. However, what I have in mind is a broader view of mathematics as a unified subject that goes back at least as far as Newton and Leibniz, and that numbers among its masters Euler, Gauss, Abel himself, Galois, Riemann, Poincaré, Hilbert, and others of past centuries.

It may only be my own prejudice as to what is first rate, but I am convinced that most first rate mathematicians have this broad view of mathematics and do read the masters. An outstanding example of this is provided by Carl Ludwig Siegel's essay on the history of the Frankfurt Mathematics Seminar [1] where he describes the participation of faculty

Harold M. Edwards is Professor of Mathematics at New York University. He received a B.A. from the University of Wisconsin in 1956, an M.A. from Columbia University in 1957, and a Ph.D. from Harvard University in 1961, and has been on the faculties of both Harvard and Columbia. Edwards's primary interests are number theory and the history of mathematics; his two recent books—*Riemann's Zeta Function* (1974) and *Fermat's Last Theorem* (1977)—for which he was awarded the Steele Prize for Mathematical Exposition of the American Mathematical Society in 1980, explain topics in number theory in terms of their historical development. He is now at work on a third book on history and number theory.

and students in Frankfurt in the 1920's and early 1930's in a seminar on the history of mathematics.

> It was the rule in those seminars to study the more important mathematical discoveries from all epochs in the original: everyone involved was expected to have studied the text at hand in advance and to be able to lead the discussion after a group reading. In this manner we studied ancient mathematicians and for many semesters devoted ourselves to a detailed study of Euclid and Archimedes. Another time we spent several semesters on the development of algebra and geometry from the Middle Ages to the mid-17th century, in the course of which we became thoroughly familiar with the works of Leonardo Pisano, Vieta, Cardano, Descartes and Desargues. Our joint study of the ideas from which infinitesimal calculus developed in the 17th century was also rewarding. Here we dealt with the discoveries of Kepler, Huygens, Stevin, Fermat, Gregory, and Barrow, among others.

To most of us, the idea of reading Leonardo Pisano (alias Fibonacci, 1180?-1250?) may seem to carry the idea of reading the masters too far, but the lasting value of the activities of this history seminar is attested to by more than Siegel's tribute to it as having afforded him "some of the happiest memories of my life."

The guiding spirit of the history seminar was Max Dehn, one of the outstanding mathematicians of the 20th century, who made fundamental contributions to combinatorial group theory, combinatorial topology, and the foundations of geometry. Dehn's work is characterized by its imaginativeness and the fertility of its ideas. Those who knew him see a direct connection between these qualities and Dehn's profound interest in the history of mathematics. His seminar dealt specifically only with mathematics before the 18th century—perhaps because the mathematics of these periods belongs equally to mathematicians in all the modern specialties, so that geometers, algebraists, analysts, topologists, and all could take part—but for Dehn and his colleagues, the method, the approach, and the belief in the value of this type of study went far beyond the 18th century.

Perhaps the greatest example of a combination of historical scholarship and mathematical research is Siegel's discovery of what are now known as the Riemann-Siegel formulas in the theory of the zeta function. Around 1930, Siegel undertook the study of Riemann's *Nachlass* in analytic number theory, the disconnected jottings of formulas on loose sheets which were in Riemann's papers when he died in 1866, and which are preserved in the Archives of the Göttingen State and University Library. Because of the great interest in the zeta function and in Riemann's hypothesis that its zeros (other than the ones at the negative even integers) all have real part equal to 1/2, there was great curiosity to know what unpublished information Riemann might have had about the zeta function. Others had tried before Siegel to make some sense out of Riemann's disconnected notes, but the task required someone who was at the same time a mathematician with great sophistication and technical ability, and an historian with the dedication and patience to analyse an original text such as Riemann's. The result of Siegel's efforts was the astonishing discovery that Riemann possessed two formulas for the zeta function, each of which was a substantial contribution to the theory of this function in 1932, seventy-five years after Riemann's death! It is entirely conceivable that, were it not for Siegel, these formulas would still be unknown today.

Another major mathematician of the 20th century who is an avowed reader of the masters is André Weil.

> As a young *normalien* I had studied Riemann, and later Fermat; I was persuaded very early that diligent attention to the great mathematicians of the past is a source of inspiration no less fertile than the reading of the fashionable authors of the day.* [2, v. 1, p. 520]

Weil visited Dehn's seminar in Frankfurt in 1926 and "as often as possible" from then on. He has written that in the sessions of the seminar "I felt I was taking part in an incomparable intellectual festival"** [2, v. 3, p. 460], and,

> having had the benefit of such an experience, I naturally found myself led, at the time Bourbaki began his works, to propose that these include historical commentaries to put in proper perspective the expositions, which were in danger of falling into excessive dogmatism. For a while, the care of these was placed mainly in my hands, and the preliminary plans of this nature that I submitted to Bourbaki were generally approved with a minimum of revisions, contrary to what always happened with the properly mathematical texts that Bourbaki received from his collaborators . . . Little by little, Bourbaki's other collaborators acquired a taste for this type of work, and my own participation became more and more sporadic . . . *** [2, v. 3, p. 460-461]

Thus, through Weil, the collaborators of Bourbaki—perhaps even Bourbaki himself—became readers of the masters.

---

* *Jeune normalien, j'avais étudié Riemann, puis Fermat; je m'étais tôt persuadé que la fréquentation assidue des grands mathématiciens du passé est une source d'inspiration non moins féconde que la lecture des auteurs à la mode du jour.*

** *. . . j'eus le sentiment de prendre part à une incomparable fête intellectuelle.*

*** *Ayant eu le bénéfice d'une pareille expérience, je me trouvai naturellement porté, lorsque Bourbaki commença ses travaux, à proposer d'y faire figurer des commentaires historiques pour replacer dans une juste perspective des exposés qui risquaient de tomber dans un dogmatisme excessif. Pendant un certain temps, c'est à moi surtout qu'en incomba le soin, et les avant-projets de cette nature que je soumis à Bourbaki furent généralement approuvés avec un minimum de retouches, contrairement à ce qui s'est toujours passé pour les rédactions proprement mathématiques que Bourbaki recevait de ses collaborateurs. . . . Peu à peu, d'autres collaborateurs de Bourbaki prirent goût a ce genre d'exposés, et ma participation devint de plus en plus sporadique . . . .*

There are several reasons that the best mathematicians are attracted to the classics. For one thing, they are, by nature, synthesizers. The essence of mathematics is the perception of logical relationships between seemingly disparate ideas; the most successful mathematicians are those whose range is widest and whose ability to see analogies and connect ideas is greatest. The role of history in such activity is obvious. Often the analogies have been recognized and developed for generations, and often the connections are most easily made via a common historical antecedent. More often than we like to imagine, advances result not from new ideas, but from the realization that old ideas apply in new circumstances.

A colleague of mine, who agreed that the best mathematicians study history, suggested yet another reason. "Good mathematicians work on good problems," he said, "and good problems have histories." He might have added that it is often the history that *makes* it a good problem, because it establishes the connections with the solutions of other problems and with applications both inside and outside mathematics.

Of course, all the usual reasons for studying history of any kind apply equally well to mathematics. History is the accumulated experience of mankind, and no intelligent person ignores it. It has shaped what we are, and it is our best source of information as to what is possible—or impossible. Countless aspects of our complicated and mystifying world are rendered intelligible only by an understanding of how they came to be as they are.

In mathematics, however, the importance of history is even greater. Mathematicians talk about mathematical "objects" and like to think that there is an objective truth to the theorems that they prove. But no one has ever seen one of these "objects" and in the last analysis the only way one can learn about them is to read about them in the works of others. For this,

---

Mathematics, like philosophy, is virtually inseparable from its history.

---

one naturally is well advised to read the works of the most clear-sighted authors, the masters. From this point of view, mathematics, like philosophy, is virtually inseparable from its history. The central concepts, the problems to be studied, and the way the theories are organized, are all inextricably associated with the names and works of the authors who first advanced them.

Finally, a simple reason for the attraction of the best mathematicians to history, which may supersede all the others, is that history is so *interesting*. First rate mathematicians are, naturally, passionately interested in mathematics, and are consequently interested in the great mathematics of the past, which is better than a major part of the mathematics of the present.

And the stimulation of interest does not go in only one direction. Many times mathematicians whose interest in their own work was flagging have found new inspiration by reading the classics.

For all these reasons, reading the masters is an important part of mathematical life and should, therefore, be an important part of mathematical education. Unfortunately, it seldom is. The main reason, no doubt, is lack of time. American graduate students need to learn so much in so little time and need to start doing research for a dissertation so soon that there is no time for the classics. In this situation, it is the responsiblity of the faculty to give students guidance, to explain that it is impossible to learn everything at once, to make them aware of the value of reading the classics, and to help them acquire the skills necessary to do so.

> *One must bring to older writings an understanding of the mathematics of the period in which they were written.*

To read the classics in mathematics is not as easy as it might sound. Changes in terminology, in style, and in the general mathematical milieu make mathematics (regarded, ironically, by the man in the street as an immutable body of facts) a field in which works written only a decade or two ago often sound archaic to young readers. One must bring to older writings an understanding of the mathematics of the period in which they were written—the assumptions their writers made about the reader as regards basic concepts, known theorems, problems to be considered, customary methods of proof, and so forth. It is not easy for students to acquire this kind of background information. Patient study is the only method. Courses in the history of mathematics cannot normally give enough attention to any one field of modern mathematics to give specialists in that field all the information they need to read the classics. It is of great help to find a thesis supervisor who has a good grasp of the classical literature in the field and who will direct the student's reading. Unfortunately, even supervisors who do know the classics well often fail to see the importance of opening them up to their students, and, even more unfortunately, there are many supervisors who have only a superficial knowledge of the classics.

Another skill that students must develop in order to read the masters is the knowledge of languages. Before World War II, very few of the masters wrote their works in English. If American students are going to study these works, they will have to study foreign languages, too. This important component of education—useful in all areas of mathematical research, not just in reading the classics—is increasingly ignored today, to the long-term disadvantage of our students. I have known young Ph.D. mathematicians who specialized in number theory, a field in which at least 80% of the classical literature is in German, and who were allowed to enter adult life without even the ability to work their way through a German mathematical text with the aid of a dictionary. They hope the classics will be translated, but this is unlikely, because the handful of people who have both the linguistic and the mathematical ability to make adequate translations are unlikely to want to spend the needed time and effort for the benefit of the handful of potential readers.

The continued health and vitality of mathematics depends on leaders in the mathematical community who are versatile and active in several areas of mathematics and who are therefore able to oppose the tendency toward fragmentation that results from there being so many specialties that are pursued so intensely. There are many such leaders today, and we owe to them several wonderful syntheses in recent years that have brought together techniques from separate specialties, thereby enhancing these techniques and increasing their applicablity. The generation now being educated will be more likely to produce such leaders if they heed Abel's advice and read the masters.

References

[1] C.L. Siegel. "Zur Geschichte des Frankfurter Mathematischen Seminars." *Frankfurter Universitätsreden*, 1964, Heft 36; also in *Gesammelte Abhandlungen*, V. 3, pp. 462-474. Authorized English translation in *The Mathematical Intelligencer* 1:4 (1979) 223-230.
[2] A. Weil. *André Weil: Oeuvres Scientifiques (Collected Papers)*. Springer-Verlag, New York, 1979.

Mathematics as Propaganda

Neal Koblitz

One night several years ago while watching TV, I was surprised to see a mathematical equation make an appearance on the Tonight Show. The occasion was an interview with Paul Ehrlich, author of *The Population Bomb* and popularizer of population control as a solution to the world's problems. At that time the ecology movement had just started to capture the attention of the public, and Mr. Ehrlich was arguing that the solution, as always, was in population control.

Johnny Carson was in top form, but the show could have bogged down if his guest had delved into subtleties or overly serious discussion. However, Ehrlich had the perfect solution. He took a piece of posterboard and wrote in large letters for the TV audience:

$$D = N \times I.$$

"In this equation," he explained, "*D* stands for damage to the environment, *N* stands for the number of people, and *I* stands for the impact of each person on the environment. This equation shows that the more people, the more pollution. We cannot control pollution without controlling the number of people."

Johnny Carson looked at the equation, scratched his head, made a remark about never having been good at math, and commented that it all looked quite impressive.

Who can argue with an equation? An equation is always exact, indisputable. Challenging someone who can support his claims with an equation is as pointless as arguing with your high school math teacher. How many of

Neal Koblitz is Associate Professor of Mathematics at the University of Washington in Seattle. After receiving his B.A. degree from Harvard in 1969, he studied algebraic geometry and number theory at Princeton where he received his Ph.D. in 1974. In 1974-75 and again in 1978 Koblitz studied at Moscow University. From 1975 to 1979 he was a Benjamin Peirce Instructor at Harvard University. He is the author of two books and several research articles on algebraic number theory and arithmetic algebraic geometry.

Johnny Carson's viewers had the sophistication necessary to question Ehrlich's equation? Is Ehrlich saying that the "*I*" for the president of Hooker Chemicals (of Love Canal notoriety) is the same as the "*I*" for you and me? Preposterous, isn't it? But what if the viewer is too intimidated by a mathematical equation to apply some common sense? Ehrlich knew how to use his time on the show well.

Political theory

Of course, it will surprise no one to find low standards of intellectual honesty on the Tonight Show.

But we find a less trivial example if we enter the hallowed halls of Harvard University, where Professor Samuel Huntington lectures on the problems of developing countries. His definitive book on the subject is *Political Order in Changing Societies* (1968), in which he suggests various relationships between certain political and sociological concepts: (a) "social mobilization," (b) "economic development," (c) "social frustration," (d) "mobility opportunities," (e) "political participation," (f) "political institutionalization," (g) "political instability." He expresses these relationships in a series of equations (p. 55):

$$\frac{\text{social mobilization}}{\text{economic development}} = \text{social frustration} \left(\frac{a}{b} = c \right)$$

$$\frac{\text{social frustration}}{\text{mobility opportunities}} = \text{political participation} \left(\frac{c}{d} = e \right)$$

$$\frac{\text{political participation}}{\text{political institutionalization}} = \text{political instability} \left(\frac{e}{f} = g \right)$$

When he is called upon to summarize his book (e.g., in *Theories of Social Change*, Daniel Bell, ed.), he emphasizes these equations.

Huntington never bothers to inform the reader in what sense these are equations. It is doubtful that any of the terms (a)-(g) can be measured and assigned a single numerical value. What are the units of measurement? Will Huntington allow us to operate with these equations using the well-known techniques of ninth grade algebra? If so, we could infer, for instance, that

$$a = b \cdot c = b \cdot d \cdot e = b \cdot d \cdot f \cdot g,$$

i.e., that "social mobilization is equal to economic development times mobility opportunities times political institutionalization times political instability!"

A woman I know was assigned an article by Huntington for her graduate seminar on historical methodology. The article summarized his work on modernization and cited these equations. When she criticized the use of the equations, pointing out the absurdities that follow if one takes them seriously, both the professors and the other graduate students de-

murred. For one, they had some difficulty following her application of ninth grade algebra. Moreover, they were not used to questioning an eminent authority figure who could argue using equations.

. . . mystification, intimidation, an impression of precision and profundity . . .

Huntington's use of equations produced effects—mystification, intimidation, an impression of precision and profundity—which were similar to those produced by Paul Ehrlich's use of an equation on the Tonight Show. But Huntington operates on a more serious level. He is no mere talk-show social scientist. When he is not teaching at Harvard, he is likely to be advising the National Security Council or writing reports for the Trilateral Commission or the Council on Foreign Relations.

Slavery

Before leaving Harvard, let us look in on another professor, this time in the Department of Economics. Robert W. Fogel's specialty is applying quantitative methods to economic history. He and a collaborator, Stanley Engerman, produced a sensation in 1974 with a book called *Time on the Cross*. Using statistical arguments with voluminous computer-processed data, they purported to show that the slave system in the South was both more humane and economically more efficient than the free labor system that existed at that time in the North.

Although this thesis contradicted the conclusions of all major conventional historians, the book was received enthusiastically. Harvard historian Stephan Thernstrom called it "quite simply the most exciting and provocative book I've read in years," and Columbia economist Peter Passel wrote in his *New York Times* book review that it has "with one stroke turned around a whole field of interpretation and exposed the frailty of history done without science."

The initial acclaim lasted long enough to produce an effect outside academia. Fogel appeared on the Today Show; the book was reviewed in the *Wall Street Journal, Time Magazine, Newsweek,* and over three dozen other major publications. The public was told that a sentimental and subjective view of slavery had given way to a "scientific" view based on computer analysis of hard quantitative facts.

But then historians of the slave period and specialists in the use of quantitative methods in history ("cliometricians") undertook careful studies of the book, and the honeymoon ended. They found such an accumulation of outright errors, fallacious inferences, dubious assumptions, and disingen-

uous use of statistics that the entire project lost any validity. Here is a
typical example, as explained by Thomas L. Haskell in the *New York
Review of Books*:

> ... readers of *Time on the Cross* are inclined toward a benign view of slavery
> when they read that the average slave on the Barrow plantation received only
> 0.7 whippings per year. In the first place the figure is too low because it is
> based on an erroneous count both of the number of slaves Barrow owned and
> the number of times he whipped them. But more important, the figure is not
> the most relevant measure of the importance of whippings. A whipping, like a
> lynching, is an instrument of social discipline intended to impress not only
> the immediate victim but all who see or hear about the event. The relevant
> question is "How often did Barrow's slaves see one of their number
> whipped?" —to which the answer is every four and a half days. Again, the
> form in which the figures are expressed controls their meaning. If one
> expressed the rate of lynchings in the same form Fogel and Engerman chose
> for whippings, it would turn out that in 1893 there were only about 0.00002
> lynchings per black per year. But obviously this way of expressing the data
> would cause the reader utterly to misunderstand the historical significance of
> the 155 Negro lynchings that occurred in 1893.e explicit.

Other examples would take too long to go into here; the interested
reader is referred to Haskell's excellent article (*NYRB*, Oct. 2, 1975) and to
the three volumes critiquing *Time on the Cross* which Haskell reviews.

Haskell regards *Time on the Cross* as an aberration, and refrains from
indicting the entire "cliometric" approach because of one unfortunate case.
However, he makes some insightful comments on the dangers inherent in
any application of mathematics to the social sciences:

> On the surface, cliometrics is an austere and rigorous discipline that mini-
> mizes the significance of any statement that cannot be reduced to a clear
> empirical test ("operationalized"). But beneath the surface one often finds
> startling flights of conjecture, so daring that even the most woolly-minded
> humanist might gasp with envy.

> The soft, licentious side of cliometrics derives, paradoxically, from its reliance
> on mathematical equations. Before the cliometrician can use his equation to
> explain the past, he must assign an empirical value to each of its terms, even
> if the relevant empirical data have not been preserved or were never recorded.
> When an incomplete historical record fails—as it often does—to supply the
> figures that the cliometrician's equations require him to have, it is considered
> fair play to resort to *estimation*, just so long as he specifies the assumptions
> underlying his estimates. And although cliometrics requires that these and all
> other assumptions be made explicit, it sets no limit at all on the *number* of
> assumptions one may make, or how high contingent assumptions may be
> piled on top of each other—just so they are explicit.

Fogel, like Huntington, understands the propaganda value of mathematics. In some quarters, invoking an equation or statistic can be even more

persuasive than citing a well-known authority. An argument which would be quickly disputed if stated in plain English will often acquire some momentum if accompanied by numbers and formulas, regardless of whether or not they are relevant or accurate. The threshhold of expertise and self-confidence needed to challenge an argument becomes much higher if it is enshrouded in science. It is no wonder that quantitative methods have become a bit of a fad in the social sciences.

The impact of *Time on the Cross* reached outside the academic world. Slavery is perhaps the most profound and emotional issue in American history. How one regards slavery has clear implications for attitudes toward present-day grievances of black people and methods proposed to address those grievances, such as busing, affirmative action, compensatory education, etc. It was because of these implications that the book received so much attention outside scholarly circles.

Another example of pseudo-quantitative arguments injected into an emotional issue with wide repercussions is the IQ controversy.

IQ

Cyril Burt is often regarded as the father of educational psychology. During a prolific research career that spanned several decades he influenced much thinking among both psychologists and educators in Great Britain and the United States. He was knighted, and his recognition in America included the Edward Lee Thorndike Award of the American Psychological Association.

One of his major achievements was his studies of identical twins separated at birth, which purported to show that intelligence is determined predominantly by heredity rather than by environment. The idea of these studies was that such twins have the same genes but are raised in different environments, so that the correlation between their performance on IQ tests gives a measure of the relative influence of heredity. Burt concluded from his studies that IQ is about 80% heredity and 20% environment.

These studies became especially well known in the late 1960's and early 1970's, because they provided the most important scientific argument for the position popularized by Berkeley Professor Jensen, Stanford Professor Shockley and Harvard Professor Herrnstein that inequalities in society are explained largely by the genetic inferiority of those on the bottom. Jensen and Shockley maintained that 4/5 of the roughly 15 point difference in the measured average IQ of blacks and whites is due to racial differences in intelligence; Herrnstein claimed that "as technology advances, the tendency to be unemployed may run in the genes of a family about as certainly as bad teeth do now." Of course, this viewpoint met vigorous opposition: scholars in various fields pointed out the inaccuracies and cultural biases of IQ tests, the effect of the testing situation, the logical fallacy of extrapolating from interpersonal to intergroup differences, and so on. But for a long time no one went back to examine in detail Burt's original studies.

The first person to do so was Princeton psychologist Leon Kamin. His curiosity aroused by the stormy controversy, Professor Kamin started reading Burt's papers and trying to locate his raw data. Almost immediately he came upon startling irregularities. For example, Burt published three reports in 1955, 1958 and 1966, during which time the total number of identical twins reared apart and the total number of identical twins reared together increased, presumably as more data came in. The reported correlation of IQ's in each group remained identical to three decimal places!

Burt's reported correlations for IQ's of identical twins

	Reared apart			Reared together	
1955	0.771	$(N_1 = 21)$		0.944	$(N_2 = 83)$
1958	0.771	$(N_1 = $ "over 30")		0.944	$(N_2$ unspecified)
1966	0.771	$(N_1 = 53)$		0.944	$(N_2 = 95)$

The chance of this happening by honest coincidence is infinitesimal. It began to look like Burt faked his data. The more Kamin examined Burt's work, the more evidence he found of fabrication. (The interested reader may consult Kamin's fascinating book *The Science and Politics of IQ* (John Wiley & Sons, 1974), from which the above chart is taken.)

Backed up by his quantitative "studies," such as the series on identical twins, Cyril Burt had an enormous impact on educational practice as well as theory.

His view that intelligence was predetermined at birth and largely unchangeable helped to shape a rigid, three-tier school system in England based on an I.Q. test given to children at the age of 11. ("Briton's Classic I.Q. Data Now Viewed as Fraudulent," *New York Times*, Nov. 28, 1976.)

More recently, Jensen has used Burt's research in a long article in the *Harvard Educational Review* (1969) to argue the futility of compensatory education programs such as Head Start.

Finally, it is amusing to note how those who are ideologically wedded to the hereditarian position reacted to the exposure of Burt. First of all, they fell back on other studies that seemed to support Burt. These studies were also analyzed by Kamin and shown to be full of holes (though not because of deliberate falsification). But it would take more than Kamin's scientific analysis to shake the self-confidence of Burt's disciples. Interviewed by the *Harvard Crimson* (Oct. 30, 1976), Herrnstein said that even "if he did fake his data, then he faked it truly."

A game anyone can play

The manipulative use of quantitative arguments is most glaring and annoying when the result is distasteful. But such methodology is merely a tool, and like most tools can be used for good or evil. The word "propaganda" in my title is not necessarily meant to be pejorative. All I mean by that word is a device that makes it possible to disseminate and popularize a point of view without a thorough and careful argument. If frightening statistics on smoking and cancer or drinking and auto accidents are presented in the high schools in a slightly over-simplified and misleading manner, then that is propaganda; but most people would feel that in such cases "the ends justify the means."

I remember using this tool once for a polemical purpose. The year was 1969, and I, together with other anti-War activists, was supporting striking workers at a General Electric plant near Trenton, New Jersey. We were attempting to build ties between workers and students, at the same time encouraging workers' increasing disillusionment with the Vietnam War. In addition to moral and material support for the strike, we brought leaflets. My contribution to one of the leaflets was the graph in Figure 1. Its message is clear: as a defense contractor, GE's profits went up after the escalation of the War, at the same time as workers were paying for the War through battle deaths and declining real wages.

My use of the graph was admittedly an example of "mathematics as propaganda." Someone unsympathetic to my purpose could complain that

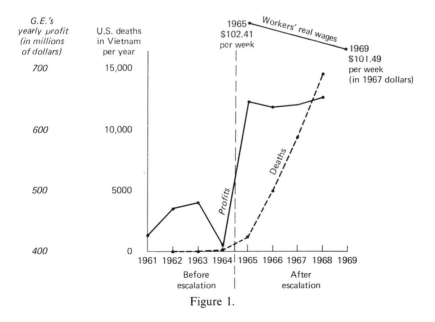

Figure 1.

the graph unfairly implies that GE profits *cause* GI deaths. I was exploiting the basic fact that the graphs of any two increasing functions can be made to resemble one another if the scales and intervals for the two variables are suitably chosen. On the other hand, someone who agreed with what we were doing would probably consider the graph to be a permissible short-cut to get the point across.

Implications for teaching

Whether mathematical devices in arguments are used for fair ends or foul, a well-educated person should be able to approach such devices critically. Indeed, for many people it may be as important to be able to identify misuses of mathematics as it is to know about the correct uses. More generally, discussions of the fallacies of the pseudo-sciences (astrology, laetrile, biorhythms, etc.) should be included as part of the basic science program in the schools.

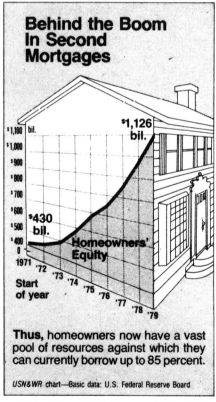

Figure 2. (With permission from *U.S. News and World Report*, Nov. 12, 1979.)

Assumption for model of world population growth	Differential equation	Solution (in billions of people)	Predicted population in 100 years (in billions of people)
(1) growth rate remains constant at 2% per year	$y' = 0.02y$	$4e^{0.02t}$	29.6
(2) an inhibiting effect proportional to the population decreases the growth rate	$y' = 0.02y\left(2 - \dfrac{y}{4}\right)$	$\dfrac{8}{1 + e^{-0.04t}}$	7.9
(3) the growth rate itself decreases by 2% per year	$y' = 0.02y\, e^{-0.02t}$	$4e^{1 - e^{-0.02t}}$	9.5

Figure 3.

Concretely, how can we impart the analytical abilities and the qualities of skepticism and sophistication needed to be able to deal intelligently with quantitative arguments about social and psychological phenomena? This is a difficult challenge. All I can hope to do here is illustrate with a couple of examples.

When students study the use of graphs, they could be shown (or asked to find) examples from the newspapers and newsmagazines which illustrate how graphs can be set up so as to exaggerate a trend or over-simplify a situation. A homework problem might be: Find at least three ways in which the graph in Figure 2 manages to convey an exaggerated impression of the data. (Answer: the vertical scale starts at 400 rather than zero, the vertical scale is larger to the right than to the left, and actual dollars, rather than inflation-adjusted dollars, are used.)

The table in Figure 3 illustrates an example I have used with first year calculus students who are just learning about differential equations. I describe three possible models for world population growth, all plausible, all based on the same empirical data (present population 4 billion, present growth rate 2% per year), and all leading to differential equations which can readily be solved by the technique of separation of variables. I then ask the students what one can conclude from the widely disparate predictions of the three models. Usually, none of them knows what to say, since mathematics is always supposed to provide a definitive precise answer. Of course, the moral of the story is that world population is far too complex to follow any simplistic model.

Mathematicians in a more public role

In addition to our influence as teachers, mathematicians can affect public attitudes through various institutions and media. Just as the medical profession tries to combat medical quackery, members of the mathematical profession can take a stand against mathematical quackery.

I know of one recent example of a remarkably effective effort in this direction by a mathematician: Serge Lang's campaign against Seymour Martin Lipset's ill-conceived "Survey of the Professoriate." Lang's comprehensive critique of the biases, improprieties and methodological fallacies in the survey has led many people to a re-examination of surveying in general and of the use of pseudo-quantitative techniques in sociology. I cannot go into Lang's critique here; the reader is referred to his article in the *New York Review of Books* (May 18, 1978) and to the volume of correspondence on this issue soon to be published by Springer-Verlag.

To stem the tide of pseudo-science and educate the public to an accurate appreciation of science and mathematics is a Sisyphean task. Major progress in this endeavor comes only at the expense of a tremendous investment of time and energy, as Leon Kamin and Serge Lang can attest.

But we can start by opening our eyes to the fact that the relationship between mathematics and society is far more extensive than merely its direct relationship to technology. Mathematical and statistical methodology is playing an increasing and even dominant role in many fields which are far removed from what one would have thought to be readily quantifiable. As we have seen, this proliferation is not an unmixed blessing. Abuse is pervasive. Mathematics can be used to mystify and intimidate rather than to enlighten the public. As mathematicians, we should take an interest in this perversion of our discipline. There is no easy solution. But the first step in solving the problem is to recognize that it exists.

Mathematicians Love Books

Walter Kaufmann-Bühler, Alice Peters, and Klaus Peters

Many of us remember certain books, often specific pages, their layout, the arrangement of definitions, theorems and proofs. These were the passages that influenced our development as mathematicians. Such experiences make us identify with yellow, green or blue books for the rest of our lives as mathematicians; they form our ideas of style and typography.

Mathematicians love books—it's the only thing they have: yardsticks of ideas, references to the known, roads towards new discoveries. There is a tremendous richness of ideas in our libraries. Often it seems indeed more than we can digest, and rediscoveries are not at all rare. There is such an abundance of results in the literature of the last 150 years that this does not come as a surprise. Each individual discovery is a testimonial to the vigor, the timelessness, and the inventiveness of the mathematical mind.

Walter Kaufmann-Bühler is Mathematics Editor of Springer-Verlag. He was born in 1944 in Heidelberg (West Germany) where he attended the local Gymnasium and University. In 1968 he received his degree (*Diplom*) under K. Jörgens for a thesis on half–bounded differential operators. Currently his main interest is historical; in 1981, Springer-Verlag will publish his first book, a biography of Gauss. Since 1968, Kaufmann-Bühler has been working for Springer-Verlag, first in Berlin and Heidelberg and, for the last 8 years, in New York.

Alice Peters is Vice President of Birkhäuser Boston, Inc., which she started in 1979 jointly with Klaus Peters and the owners of Birkhäuser Verlag Basel, an old Swiss book publisher specializing in science, art, history and architecture. She received a B.S. in mathematics in 1967 from the University of Rochester, and an M.S. in computer science at the University of Chicago in 1969. In 1971 she became mathematics editor of Springer-Verlag Berlin-Heidelberg.

Klaus Peters is President of Birkhäuser Boston, Inc., which he started in 1979 jointly with Alice Peters and the owners of Birkhäuser Verlag Basel. Born in Wuppertal, Germany, he studied in Munster and in Erlangen, where he received his Ph.D. in 1962 under Reinhold Remmert. After teaching at the University of Erlangen for two years, he became mathematics editor at Springer-Verlag in 1964, and one of the five directors of Springer-Verlag in 1974. He is also President of Suhrkamp/Insel Publishers Boston, Inc., an affiliate of the German literary publisher.

The authors of this article have all at one point in their career decided to "give up" mathematics in favor of mathematical publishing. We hope that the account of our understanding of (mathematical) publishing, tinted as it may be by enthusiasm, and some vested interest, will cast light on motivations and decisions about which many mathematicians may have wondered.

Mathematics publishing, like other publishing, grew out of the need and desire to communicate and—quite a different aspect—to preserve knowledge for one's own and for the community's sake. Over the years there have been very different attitudes towards publishing, strongly influenced by the social environment and by the demands and criteria imposed upon mathematicians. While Gauss was satisfied to produce much of his work for his notebooks and to communicate selected morsels in his extensive correspondence, mathematicians of the 20th century have been pushed into a *publish* or *perish* situation. This juxtaposition contains an obvious argument that runs counter to the "social influence": the quality of the mathematician is reciprocal to his need to publish. We suppose that this inner contradiction has caused a mathematician-publisher to name his company "Publish or Perish" and to choose Gauss's motto for his advanced book series: "*Pauca sed matura.*"

But there is another side to this. Perhaps mathematics is getting old, and today's discovery might be the last. Moreover, the age of the book as a vehicle for scientific information may approach its end: in the not too distant future, mathematics may no longer be studied in the traditional way, nor transmitted on printed paper between two covers. Will the book as we know it survive? Our emotional responses might be mixed.

The various functions of a publisher as they emerged over centuries can be broken down into *acquisition*, which presumes a certain amount of market research and critical evaluation, *production*, which includes copy-editing, design, typesetting, printing and binding, *warehousing and distribution*, which is based on promotion and sales activities, and *financing*, which stands at the beginning and end of all business operations. In fact, while the English word "publisher" emphasizes the purpose, the German word "*Verleger*" points to the fact that the publisher advances the money (*vorlegen*) to make publication possible. The history of most publishers who have remained in business over an extended period of time supports the statement that an initial personal interest of just one person, the publisher, led to the development of a program and determined the subsequent policy. This does not necessarily mean that the publisher must be an expert. His personal interests and friendships, however, further and expand the program that his company represents.

In the selection process for scientific books, negative decisions can often be made on the basis of a general acquaintance with the subject and on stylistic or economic grounds. Positive decisions are almost invariably based on peer review, and as such, necessitate strong ties between the publisher and the scientific community. While in the beginning of this century it was a rather well-defined question whom to invite as editor for a specific discipline, the growth of mathematics, the diversification of subject areas and the multiplication of scientific centers in a geographically expanding community has added to the choice, for better or for worse. Not being able to rely on the advice of one or two undisputed authorities in the field, the in-house editor needs to exercise his own scientific judgment, not necessarily to evaluate the scientific quality of a manuscript, but to select the right expert, eliminate marginal proposals, and develop new publishing ideas as a result of discussions with potential authors and readers.

In the beginning, publishing grew out of printing. Those who mastered the art of moving letters possessed a power and exerted an influence often in alliance with the spiritual leaders of their time. Traditionally, typesetting has played a major role in mathematics publishing. Often a relationship of mutual respect has developed between authors and typesetters who through years of experience were able to discover mathematical inconsistencies by recognizing deviations in the structure of formulas.

Sophisticated word processors equipped with programs to handle mathematical text are available already . . . .

Most mathematicians now understand that growing economic pressures have dictated the use of camera-ready manuscripts for specialized monographs in order to avoid a pricing policy that would impede dissemination of knowledge. Some recent developments and the interest with which they are greeted indicate, however, a potentially significant change. Sophisticated word processors equipped with programs to handle mathematical text are already available in many computer science departments. For anyone accustomed to using a typewriter in the preparation of his manuscript, such word processors add a completely new dimension. The author can type the text and, by using simple commands, insert formulas, certain graphic material, and even tables that use the computing facilities of the word processing equipment. He or she can make corrections on the computer screen without erasing, cutting, pasting, or worrying about new pagination. The author can store the complete manuscript, or several for that matter, on the word processor and recall it at any time to make changes, additions or deletions. The author can print out part of the manuscript, or all of it, to send to reviewers, colleagues, publishers, or just to take it home. At any moment the author has an up-to-date manuscript.

There are of course a number of problems still to be solved, among them the compatability of the typesetting hardware and the availability of fonts and special symbols the author may want to use. However, it seems clear that in a few years, almost everyone will have access to such typesetting facilities and will be able to present the publisher with printed camera-ready copy, or magtape or a floppy disk containing his manuscript. The question may arise as to how copyediting and reviewing will be incorporated into the process, but obviously changes in such a "manuscript" can be more easily made than on a copy typed on paper. Nevertheless, the published result of this process will not look too different from what we are already accustomed to, namely, bound books to be placed in libraries, on bookshelves or carried in a briefcase. In fact, such a development is well within the conventional framework of traditional publishing. It is even one of the objectives of computerized typesetting to regain the possibilities and elegance of 19th century methods that were gradually lost in the transition to monotype and monophoto.

Imagine . . . a machine that looks like a page of a book . . .- like one of the small calculators that fits in your wallet . . . with a touch-tone dial which allows the reader to select any page of any book.

It is not inconceivable that a very different step will be taken during the next decades. Presently the use of microfiche has been received without much enthusiasm, for obvious reasons: reading quality, clumsy equipment and retrieval difficulties stand in the way of efficient, anthropomorphic reading.

But imagine the following: a machine that looks like a page of a book, possibly a bit larger, but very thin, like one of the small calculators that fits in your wallet. This "one-page book" will have a reading area and a touch-tone dial that allows the reader to select any page of any book by simply dialing the International Standard Book Number (ISBN) followed by the page number desired. The page requested will appear on the "screen," and the reader will be able to read continuously forward, reverse, skip pages or even jump from one book to another. In the back of this one-page book is a supply of, say, ten blank pages; by pushing the appropriate code, the text on the reading screen will be copied onto one of these pages so that it may be sent to a colleague, perhaps with additional notes, or simply kept for future reference. If you smile and think that such ideas are too unrealistic, just remember the development of pocket calculators, home computers, and word processing equipment during the last thirty years, and remember to talk with us about it in ten years.

This step would obviously do away with our beloved yellow, green or blue covers, but to speculate about it may not merely be *"jeux d'esprit."* Even if it were, mathematicians should keep in mind that, to many, mathematics itself is *jeux d'esprit*—there is some mysterious connection between mathematics, computers, and technological (scientific) progress. Those beautiful curves, however, may not survive.

It does not come as a surprise that much of the inspiration for technological progress has come and will come from the mathematics community itself. There is one trivial reason (with non-trivial ramifications): computer science is a close relative of mathematics, and for computer scientists the development of an efficient, pleasing, and intuitively accessible program for mathematical composition is an interesting and rewarding problem. What may be less obvious to the outsider is the increasing identification of author, publisher and buying public. These three groups may never have been as far apart as one might think, but they undeniably are moving closer together. The step from producing the manuscript for a lecture note to printing the book is very small indeed, but then the author might find it difficult to store and distribute the fruits of his progress.

But there will always be changes. The next step on our arduous and stony road to utopia, after solving the composition problem, will be the transmission problem. Once this is solved, though we don't yet know how, the distribution of research announcements and shorter papers will be trivial. We have attempted a description of one option for distribution, and most certainly there will be others whose germs may not even be visible today, although most likely they are already here.

From all this, there is one clear and wholesome conclusion. If publishers are to survive they must look for some other *raison d'etre*: production and distribution are not enough. Such a reason, though not always remembered or recognized, does actually exist. Authors implicitly account for it in their choice of a publisher, libraries in their acquisition policies. The reading public has a certain conception of individual publishers and distinguishes quite easily between university presses and commercial publishers, less so among the various commercial ones, but they too are classified. The essential criterion is the selection process—how a publisher arrives at his decision to accept or reject a book. The traditional, scientific commercial publisher has a program. Each manuscript, before being accepted or rejected, is discussed with advisors and reviewed by specialists. The publisher will make the final decision, a commercial, even a formal judgment on the surface, but in reality of much more significance. If a publisher reflects upon his role, the result will be both elating and sobering, for he represents just another, indigenous, function of the mathematical community, dependent upon the confidence mathematicians themselves have in him, his reliability, and his efficiency in the fulfillment of his task. A publisher can easily lose respect if his refereeing policies become less strict, if his advisors are not impartial or balanced.

These thoughts are not meant to supply ideological ammunition *ad editoris majorem gloriam*. There is a direct relevance to the publisher's future role. The easier it becomes to produce and distribute a book outside of the traditional channels, the more important and distinct a role this germane function of publishing will attain: to select impartially, to give advice concerning changes in a manuscript, to lend, to whatever degree this is possible, an air of objectivity to the process of submitting intellectual products to the judgment of the community.

And as such, things will not change abruptly for better or for worse. At least for a few decades, books seem to be here to stay: there is no alternative in sight for that substantial part of mathematics that consists of developing well-rounded coherent mathematical theories. That is an essential part of good mathematics for which the physical object, a bound book, is the most becoming. Books will continue to be produced by publishers, faced, at the same time, by a task both easier and more difficult than their present one. There will be less distraction by the indispensible paraphernalia of publishing and more recognition of the publisher's real job, not that of arbiters of scientific development, but of honest agents of the community as a whole.

A Faculty in Limbo

Donald J. Albers

Have our still very young two-year colleges changed too rapidly during the last 20 years? Could it be that their remarkably fast enrollment increases and greatly broadened program focuses have produced change faster than most faculty can cope with? Has this growth and change been so swift that mathematics faculty of two-year colleges are in limbo and becoming recluses?

In this paper, we first chronicle the rapid growth of two-year colleges, discussing changes in their programs and student populations; then we draw a profile of today's faculty, and try to understand their role in the development of two-year colleges. (The best sources of information on the mathematics faculty and programs of two-year colleges, including more data than one probably desires, can be found in [1], [2], and [3].)

Explosive growth

During the last 20 years, no other sector of higher education has grown so rapidly as have two-year colleges. Nationally, their enrollments more than tripled in the 60's; in the 70's, they doubled. As for the 80's, two-year colleges are the only postsecondary institutions expected to grow.

In 1960, two-year colleges accounted for only 15% of all undergraduate enrollments in mathematics. Today, only 20 years later, they constitute nearly 40% of all enrollments.

Donald J. Albers is Chairman of the Department of Mathematics at Menlo College. He attended a two-year college as a freshman, completed his B.A. in 1964 at the University of North Dakota and his M.A. in 1967 at Andrews University. Albers has taught in both public and private two-year colleges since 1968. He is a member of Phi Beta Kappa and has twice won teaching prizes at Menlo College. He is currently Editor of the *Two-Year College Mathematics Journal* and is past Chairman of the Committee on Two-Year Colleges of the Mathematical Association of America.

Influence of high-school teachers

Explosive growth of such proportions created a large number of jobs for mathematicians. Indeed, so acute was the need that teacher shortages occurred in the late 50's and early 60's. The shortages were so severe that many high-school teachers, wishing to move up the educational ladder, took advantage of Sputnik-inspired National Science Foundation institute programs and earned master's degrees. Although the number of high-school teachers moving up to two-year colleges has now declined, they still account for nearly two-thirds of all two-year college mathematics teachers. Many of them in the 60's and early 70's entered newly formed two-year colleges, most of which had a classical junior college orientation, i.e., they were patterned after the first two years of a four-year college. Most of their students were 18- and 19-year olds, planning to move on to four-year transfer colleges. The vast majority of the new faculty of the 60's and early 70's left behind grueling, 25-hour teaching loads, marked by growing discipline problems in the classroom. They were part of what was hailed as an exciting boom; and to top it off, they had the added status of being college teachers.

Teaching versus scholarship

The new colleges of this period were often administered as part of secondary school districts. As a consequence, their faculties were composed largely of "promoted" high-school teachers. This practice of hiring from high school ranks gave a clear signal to four-year colleges that they (two-year colleges) placed their greatest value on teaching and not on the time-honored, four-year college tradition of scholarship and teaching. Thus, a separation between faculties of two-year colleges and four-year colleges was established early in their modern history.

The two-year college emphasis on teaching over scholarship remains much in evidence today. Two-year faculty retain considerable pride in their ability to teach and are generally hesitant to hire a Ph.D. in mathematics, who has been prepared to do research. As further evidence of an almost anti-university attitude, one-fourth of all two-year college teachers belong to *no* professional organization. Four-year college faculty observe with pleasure that the percentage of Ph.D.'s in two-year colleges has risen from 4% in 1970 to 15% today. It is important to note however, that most of the new doctorates come from two-year college ranks, typically earned by faculty after several years on the job, and are *not* degrees in mathematics.

Changing programs

The dynamic growth of two-year colleges was accompanied by changes in their programs and student populations. These changes have been nothing short of revolutionary, causing some to wonder what the word, "college," means in the name, "community college."

In the early 60's, most two-year colleges had a liberal arts orientation, serving as feeders for four-year colleges. By the mid-60's, program emphases had undergone considerable change. A host of new programs in vocational/technical areas were introduced: data processing, dental hygiene, electronics, practical nursing, automotive mechanics, accounting, printing, bricklaying, carpentry, and police and fire science, to name a few. Today, only 40% of two-year college students are enrolled in college transfer programs. The growing majority of students are now enrolled in vocational/technical programs.

> *Current estimates suggest that 20% of all mathematics courses in two-year colleges are given by departments outside of mathematics departments.*

The mathematical needs of vocational/technical students are often different from those of the liberal arts transfer student. Subjects frequently taken by these students are remedial courses in arithmetic and elementary algebra, and hybrid courses in technical mathematics. Since enrollments in traditional college transfer courses also were growing rapidly during the 60's and early 70's, and since few faculty had any formal training as preparation for teaching arithmetic or technical mathematics, most faculty had little inclination to teach such courses to students outside the familiar junior college, liberal arts track. Many departments outside mathematics responded by developing their own mathematics courses. Current estimates suggest that 20% of all mathematics courses in two-year colleges are given by departments outside mathematics departments. The net result has been a general isolation of mathematics teachers from vocational teachers, probably to the detriment of both groups.

The philosophical differences between mathematics faculty and their administrations became more pronounced during this period of program expansion, for administrators tended to view the development of applications-oriented vocational/technical programs as the very embodiment of the term "community college," while liberal–arts–trained mathematics faculty frequently exhibited little enthusiasm and occasional scorn for new or expanded vocationally-oriented programs, which they regarded as inherently non-academic. Thus, the isolation of mathematics faculty worsened: now not only had they grown apart from vocational/technical faculty, they were also isolated from their administrations.

Impact of part-timers: two views

The isolation of mathematics faculty from administration has been aggravated by the administration's increased use of part-time faculty. In 1970, only 10 years ago, 13% of all mathematics classes were taught by part-timers; today, the figure is 30%. Since full-timers usually teach three times as many courses as do part-timers, this means, in terms of headcounts, that part-time faculty now account for more than one-half of all mathematics faculty in two-year colleges. Part-timers often are hired at the last minute to teach classes at hours when most of the full-timers have left for the day. Thus, they have little or no contact with full-timers and develop questionable commitment to the institution. Their educational qualifications are generally lower than those of full-timers. Not surprisingly, they are less likely to be members of professional organizations, less likely to be concerned with curriculum, and tend to read fewer scholarly journals.

Yet, year after year, and in spite of the cries of full-timers, the flow of part-timers increases. The administration tends to view the situation hard-headedly, asking (usually rhetorically), "What can full-time faculty do that part-timers cannot?" Full-timers, in their own defense, point out that part-timers are generally less experienced, read less about mathematics and curriculum, are less committed to the institution, and as a result, are apt to contribute far less to the development of intradepartmental cohesion and chemistry. More often than not, the administration seems more interested in whether they are able to teach their classes and get their grades in on time. Administrators are also fully aware of the substantial cost savings realized by using part-timers. Such economies can be, and have been, used to develop new programs; lately, these have usually been vocational programs. Thus, the rise of part-timers has dealt a very hard blow to full-timers. It has contributed to the growth of what they perceive as a relatively foreign, non-academic sector. It has eroded departmental morale. It has made full-timers suspicious of their administrations, often making them feel hopelessly embattled, powerless, and isolated! As a result, it has reduced their involvement in shaping directions for their institutions and forced them to seek refuge in their classrooms and sometimes, regrettably, away from their colleges.

Open-doors and changing student populations

The drift toward isolation was accelerated by changes in student populations. By the 60's, most two-year colleges had relaxed admission standards to the point that they became known as open-door institutions. Open-doors gave opportunities to vast numbers of individuals who otherwise might not have been able to obtain training that could lead to a four-year degree or to a vocational diploma. By any measure, the concept of open-doors was bold and imaginative educational engineering. It set two-year colleges apart from other postsecondary institutions and established them as bright spots in the educational firmament. There were, of course, many critics of open-doors. These critics frequently viewed the new community college idea from a narrow, liberal arts perspective, which included traditional admissions standards. Given the liberal arts orientation of most mathematics teachers, it's not surprising that open-doors represented a not-always-welcome modification of their colleges.

> *Virtually, every major element in the history of two-year college mathematics teachers suggests that they are becoming more and more isolated: from teachers of mathematics at all other levels, from their own vocational faculty, and from administrators, as well.*

The composition of the student population evolved quickly. In 1960, it was made up primarily of 18- and 19-year old high school graduates, mostly single, mostly white, mostly males, and mostly attending on a full-time basis. Today, two-thirds of the students are over 21, one-third are married, some lack high school degrees, one-fourth are minority students, and more than one-half are women. Nearly two-thirds of these students are attending on a part-time basis, and one-half start their studies after age 21.

Many of these students require training in such remedial courses as basic arithmetic and elementary algebra. The growth of remedial courses was dramatic; today, they account for 40% of all two-year college enrollments. Simultaneously, calculus enrollments have dropped to only 10% of all enrollments. This shift toward an enlarged remediation role tended to increase both the academic distance between two-year and four-year colleges and the remoteness of two-year from four-year faculty.

Virtually, every major element in the history of two-year college mathematics teachers suggests that they are becoming more and more isolated: from teachers of mathematics at all other levels, from their own vocational faculty, and from administrators, as well. Not surprisingly, during this past decade, much discussion among two-year college teachers has reduced to two simple questions: "Who are we?" and "Where are we going?"

A new organization, The American Mathematical Association of Two-Year Colleges (AMATYC), has been asking these two questions, and answering them with some success. AMATYC's membership is now 700, about 12% of all full-time, two-year college mathematics teachers. AMATYC's efforts to serve the needs of two-year college faculty has led them to establish a new journal, *The AMATYC Review.*

Long established organizations, such as the Mathematical Association of

America (MAA), the principal organization of college mathematics teachers, and the National Council of Teachers of Mathematics (NCTM), the main high school mathematics organization, also have exhibited strong interest in two-year college teachers. Their interest stems from the large share, 40%, of undergraduate enrollments commanded by this group of teachers. Both organizations have worked hard to expand their respective foci to encompass two-year college concerns. For example, the NCTM has added special two-year college sections for meetings. The MAA has also broadened annual meeting programs to encourage two-year college participation. The MAA has also established a vigorous Committee on Two-Year Colleges, and publishes a growing journal, *The Two-Year College Mathematics Journal*, which now has nearly 7,000 subscribers.

Certainly, the efforts of these organizations should benefit two-year college teachers, but one additional problem needs to be factored into the situation. It is the simple fact that, as a whole, the full-time faculty is now middle-aged and is apt to remain relatively static over the next twenty years. Current funding practices suggest increased use of part-time faculty to replace full-time teachers who resign or retire. Thus, the infusion of new, young blood is likely to occur at very low levels until the year 2000. It may well be that today's middle-aged faculty is set in its ways. If so, the next decade could prove a very frustrating one for them.

What's to be done?

Two-year colleges have evolved with such great speed that it's unlikely that even the best educational seers of the 50's could have predicted their present state. One thing that can be done today is to take time to look back 20 years and attempt to understand what has happened. It should be clear by now that two-year colleges are not junior colleges and that attempts to remake them into junior colleges will meet with frustration. Mathematics faculty can and must ask how their special skills can be put to greater advantage in discovering and implementing community needs. First steps in this direction might center on establishing links between mathematics faculty and vocational faculty. Additional useful steps involve strengthening existing bonds between two-year teachers and teachers in four-year colleges, secondary schools, and elementary schools. The isolation may be lessened if four-year faculty recognize that two-year colleges are not junior colleges, and that they may be able to learn much from two-year colleges. By the same token, two-year faculty have much to learn from their colleagues in four-year institutions and secondary schools. We must and can work together in a community-based context toward discovering who we are and where we wish to go.

References

[1]   James T. Fey, Donald J. Albers, and John Jewett. *Undergraduate Mathematical Sciences in Universities, Four-Year Colleges, and Two-Year Colleges, 1975-76.* Conference Board of the Mathematical Sciences, Washington, 1976.
[2]   Robert McKelvey, Donald J. Albers, Shlomo Libeskind, and Don Loftsgaarden. *An Inquiry into the Graduate Training Needs of Two-Year College Teachers of Mathematics.* Rocky Mountain Mathematics Consortium, University of Montana, Missoula, 1979. (A long summary of this report appeared in the March 1979 issue of the *Two-Year College Mathematics Journal.*)
[3]   Fontelle Gilbert. *Fact Sheets on Two-Year Colleges.* American Association of Community and Junior Colleges, Washington, 1979.

Junior's All Grown Up Now

George M. Miller

Exactly ten years ago this month I was helping to write an editorial for the *Journal* of the New York State Mathematics Association of Two-Year Colleges on essentially the same topic as this article. Entitled "University Dominated," the editorial expressed the frustration of a growing number of junior or two-year college mathematics educators with the lack of an honest effort by established associations (The Mathematical Association of America and the National Council of Teachers of Mathematics) to recognize and deal with the unique problems of two-year college mathematics teachers.

Junior or two-year colleges developed to fill a particular need and have become—in 1981—responsible for the education of half of all students in higher education. Yet the unique role of the two-year college may be somewhat unclear to the lay person. Generally two-year colleges are publicly-supported and locally attended. Since many of their students have weaker educational and lower socio-economic backgrounds than students at four-year colleges, two-year colleges offer extensive programs in remediation. They also offer various two-year terminal programs, and programs and courses which are highly transferable. For these reasons, the emphasis at two-year colleges is on teaching rather than on research, one of the main focuses at four-year colleges and universities.

During the 60's and early 70's two-year colleges were given the primary burden for dealing with the concept of open enrollment in higher education. Not only was a high quality college-level education to be presented to

George M. Miller is a Professor in the Department of Mathematics/Statistics/Computer Processing at Nassau Community College on Long Island. He received an A.A. degree in 1965 from Nassau Community College, a B.A. in 1969 and an M.S. in 1971 from Adelphi University. Miller was a founding member of the New York State Mathematics Association of Two-Year Colleges, and was co-founder and is presently editor of The Mathematics Associations of Two-Year Colleges Journal. He is a recipient of the State University of New York Award for Excellence in Teaching, and the award for Outstanding Contributions to Mathematics Education of the American Mathematical Association of Two-Year Colleges.

a large group of students who were poorly-prepared socially and academically, but very few faculty, at that time, had formal training in remediation for college students. These problems, combined with those of limited funds, were further compounded by curriculums that were often designed for academically superior students.

With all of these problems, two-year college educators in 1970 found existing forums in higher education for discussion and evaluation of two-year college needs to be university dominated. Federal funds to help solve these problems were usually denied. The system of evaluation of proposals, at that time, was not conducive to a full understanding of two-year college needs. For instance, the National Science Foundation peer review process for evaluation of faculty grant proposals consisted mainly of university reviewers without the necessary experience to fairly evaluate funding proposals for two-year colleges.

More than ten years ago the Committee on Undergraduate Programs in Mathematics (CUPM) analysed and made strong recommendations for undergraduate mathematics. Not only was the overall committee university dominated, but even the special *Two-Year College Panel* consisted of a majority of senior college and university people. The disproportionately low committee representation for two-year colleges which existed a decade ago sometimes continues to be the case today. I presently sit on a 14-member National Science Foundation committee charged with providing classroom materials for undergraduate students. Although the two-year colleges presently educate over half of all undergraduate students, I am the only two-year college person on this committee.

The Ph.D. with 20 published papers in abstract ring theory, although most impressive, is not by these credentials the best authority in providing classroom materials for two-year college algebra and trigonometry courses.

The argument that it is necessary to have university faculty sit on two-year college committees for the design of the transfer curriculum is justifiable. Yet this does not imply that a committee should be made up of a majority of university faculty! The Ph.D. with 20 published papers in abstract ring theory, although most impressive, is not by these credentials the best authority in providing classroom materials for two-year college algebra and trigonometry courses.

The two-year college teacher is a specialist with his or her own unique skills, providing a valuble and integral part of the educational process. Just as the grade school teacher is not practicing to be a high school teacher, the two-year college teacher is not aspiring to teach in the university; nor does he or she need university educators for leadership in his or her own specialty.

It is interesting to think that the faculty of four-year colleges and universities have traditionally been quite willing to embrace liberal causes that seek fair and proportional representation by emerging groups. Yet there is an obvious reluctance to refer decisions regarding two-year college matters to two-year college faculty.

Ten years ago appropriate graduate level educational opportunities for two-year college faculty were almost non-existent. An editorial in a national two-year college mathematics education journal in 1970 read, "A thorn in the sides of most mathematics teachers at the two-year college level has been the artificial pursuance of a terminal research-oriented Ph.D. degree in mathematics. The strange dichotomy inherent in this situation is that the two-year college teacher is primarily concerned with excellence in teaching mathematics and not the intensive research orientation emphasized by Ph.D. programs . . . ." This editorial went on to support a doctoral level Specialist in College Teaching Degree (SCT) as was attempted (unsuccessfully) by Eastern Illinois University in 1968-69.

A number of university people have understood the unique needs of the two-year college educator. In 1965 Theodore Sizer, Dean of the Graduate School of Education at Harvard, advocated in a speech given at Johns Hopkins University, a Doctor of Arts in Teaching Degree for teachers of undergraduates. In 1970 Carnegie-Mellon University offered a Doctor of Arts Degree "to prepare the candidate for a rewarding career of communications, rather than originating mathematical knowledge (research)." Over this past decade a number of graduate schools have responded with doctoral programs that are more appropriate. For example, in 1974 Adelphi University began a unique Ph.D. program called the Teaching Option "geared to the needs of college faculty of mathematics whose primary responsibility is teaching . . . ." Two-year college educators have also, in some areas, included a second masters degree in a related field as the equivalent of a non-research doctorate degree. In the area of graduate education, positive steps have been taken to help two-year college educators achieve their goals.

University dominance of two-year college committees and journals has not been limited to mathematics. The president's message in the first newsletter of *The Teachers of Accounting in Two-Year Colleges* put it this way:

> From the inception of the two-year college movement, teachers of accounting in the two-year schools have been frustrated in their attempts to communicate with their colleagues at similar institutions. We have flocked in large numbers to the AICPA (American Institute of Certified Public Accountants), the state societies, and the AAA (American Accountants' Association), only to find that the first two serve only the profession and the AAA, though educationally oriented to some degree, strongly emphasized teaching at the four year level.

Unable to effect changes within existing associations, in 1974 a group of teachers of accounting in two-year colleges founded their own association. Also in 1974, with the same motivation, a group of two-year college mathematics teachers created the framework for their own professional association, *The American Mathematical Association of Two-Year Colleges*; its membership is now nearly 1000. Both of these associations were created and still exist in response to the lack of a full partnership role for two-year college educators in existing forums in higher education.

In 1981 there is more support and understanding of the two-year college role in higher education than there was ten years ago. A number of positive changes have occurred: graduate schools have responded with a number of doctoral programs that recognize the needs of two-year college faculty; there is now a new national and at least a dozen state-wide professional associations specifically for two-year college mathematics educators; three two-year college mathematics education journals have been developed; two-year college curriculums have been modified and new programs added to successfully educate the academically weak students as well as the academically strong; and many of the public have had a first hand opportunity to experience and enjoy the benefits of the high quality of a two-year college education. But, four-year college and university dominance of higher education is still not completely gone.

One last necessary step must be taken if two-year college educators are to become *full partners* in higher education. Representation on all federally funded committees, subject matter panels, and federal and state systems for peer review of faculty grant proposals should be directly proportional to the amount of responsibility that two-year colleges hold in higher education. Specifically, this means that when the main concern is two-year college education, then the majority of the decision making body should be two-year college educators; and when the main concern involves the broader scope of all undergraduate mathematics education, then fully one-half of the decision making body should be two-year college educators.

With the development and tremendous growth of two-year colleges, higher education has changed. Had the four-year college and university system of 30 years ago been sufficient for the needs of today's society, the present two-year college system would not have emerged. Now that the two-year colleges have developed into a valuable and integral part of the process of higher education, two-year college educators will not be satisfied with token representation. The four-year college and university faculty who still resist sharing the decision-making power will, we hope, soon begin to refer two-year college matters to two-year college faculty. We can then all concentrate our efforts on the many new challenges within the system of higher education.

NSF Support for Mathematics Education

E.P. Miles, Jr.

The National Science Foundation Act of 1950 created a new Federal agency with two principal missions: providing needed support for basic scientific research and increasing both the quality of instruction in science and the number of scientists trained. The first mission was to encourage broad classes of fundamental research comparable to that which various agencies had used to accomplish military objectives in World War II, which ended five years earlier. War experience had shown the frequent dependence of applied research and development upon previous basic research often done by scientists or mathematicians when no definite applications were apparent. The mission in science education resulted in part from the rapid growth of education following the return of the veterans and the support of the GI bill. This growth was imposed upon a base of depleted faculty, many of very marginal qualifications, and outmoded equipment and curricula in the sciences. Also the booming postwar economy and rapid advances in science and technology had accelerated the need for well trained scientists and engineers.

Pre-NSF observations

To understand NSF's role in mathematics education one needs to examine the deterioration in the period 1938-1950 which led to the 1950 crisis in

E.P. Miles, Jr., is a Professor in the Department of Mathematics and Computer Science at Florida State University. His degrees, all in mathematics, were from Birmingham Southern (B.S., 1937) and Duke (M.A., 1939 and Ph.D., 1949). He taught for nine years at Auburn and twenty-two years at Florida State University, where he served ten years as Director of the Computing Center. His publications include papers on mathematics research, mathematics education, computer-related curriculum development and color graphics. He represents the Association for Computing Machinery on the Conference Board of Mathematical Sciences, where he chairs the Task Group on Science Education Funding. He is a national lecturer for the Association for Computing Machinery, speaking frequently on computer-generated color graphs of mathematical functions.

education with which NSF was created to deal. The conditions leading to the founding of NSF built up over many years. The seeds were there twelve years earlier in 1938 when, at 19, I began the full time teaching of high school mathematics. By peer standards I was then over qualified (B.S. and M.A. in mathematics plus all education courses for a Grade "A" Certificate) and by all standards underpaid ($1134 total for 9 months, a salary indicative of the pay levels that were partly responsible for the lack of adequate mathematics instructors). However, teaching mathematics was my only career goal so I stuck with it except for further graduate study or necessary military service.

A question asked of me in a class in 1938 initiated my continuing interest in improving programs for training teachers of mathematics and in modernizing curricula: "If these numbers are imaginary, why are we wasting our time with them?" My study of differential equations and complex variables provided some real uses to cite for imaginary numbers in physics, engineering or applied mathematics, but that background was rarely found in the teachers of that day. The education requirements for certification did not provide such answers. University level instruction, which I began at North Carolina State in 1940 after another year of graduate study, reinforced this interest and concern. Two examples of pre-calculus topics usually available at the high school level are cases in point: logarithms and trigonometric functions. Students who repeated these topics at the college level or who met them without repetition in the calculus sequence were by and large so locked in their thinking to only base 10 logarithms and degree-measure angles that the theory of exponents underlying other bases for logs and analytic properties of trig functions were usually missing in their backgrounds. Teachers not aware of the importance of base e exponential functions and radian measure in the development of the fundamental formulae for calculus had little motivation to cover such topics even if they were treated in their high school texts.

After five and a half years of wartime service and two and a half years needed to complete delayed doctoral studies, I resumed teaching at Auburn in the summer of 1949. By then the GI boom had caused massive growth in higher education enrollments at a time of great deterioration in the quality of mathematics instruction at pre-college and college levels during the war and early post-war years.

Some of the factors for the decline may be readily stated. Many of the better young male teachers had gone off to war where their mathematical skills declined with disuse. Other competent female or older male teachers moved into better paying industry jobs created by the war and fewer replacements were available from reduced-enrollment graduating classes. Soon after the war, waves of veteran students arrived on the nation's campuses to complete long delayed college educations or to initiate previously unplanned studies made possible by their GI benefits. This created unprecedented demands for mathematics instructors in the colleges with much of the demand being associated with remedial classes at secondary school levels. In desperation many high school teachers, some with submarginal qualifications, were hired as faculty to cope with this glut of students. This drained further the ranks of competent high school teachers and diluted the quality of college faculties.

In some cases faculty of limited training or ability slipped into tenured positions under the pressure for live bodies. For example, there was the history teacher in a rural high school whose principal announced in the late spring that he would be assigned to teach senior mathematics that fall because he was the only one who had any college level mathematics credits on the surviving faculty—in his case a courtesy "D" in college algebra more than two decades earlier. Bravely he signed up to audit algebra and to take trigonometry in the intervening summer, but, even with many help sessions from his instructor and the paid services of a competent tutor, he was unable to pass the elementary course he needed for his new assignment. As a second example consider the college instructor recruited from a high school in the late days of the war who was about to be granted tenure automatically until a faculty committee noted that his transcript showed a weak undergraduate major some 20 years earlier and no intermediate study of mathematics. The committee voted to defer his tenure decision until he could complete the normal master's degree program with fellowship and

---

His resignation . . . followed quickly when he failed the intermediate calculus course he had taught the year before.

---

teaching assistant support. He accepted this challenge with the stipulation that he be allowed first to take a few undergraduate prerequisite courses to prepare for the graduate work expected. His resignation to accept a civil service job followed quickly when he failed the intermediate calculus course he had taught the year before.

Those too young to have seen those days might conclude that all was bleak for mathematics education when the National Science Foundation was chartered and getting underway. Yet some good instruction was taking place. Many retired military officers or middle aged housewives with good mathematical training in their school years and willingness to brush up entered the teaching field and made significant contributions, as did a nucleus of dedicated professors and teachers who continued to improve and update mathematics instruction at all levels. Nevertheless, the needs to upgrade the content level of training programs for in-service teachers and to improve the pre-service curriculum for teachers in training were obviously overwhelming. The mechanisms and resources for meeting those needs on a nation-wide scale were woefully inadequate until the National

Science Foundation was established and, in its immediate post-Sputnik period a few years later, funded at a level adequate for maximal impact on science and mathematics instruction.

In the early fifties I became involved, as did many other mathematics educators, in planning new graduate courses for mathematics teachers in analysis, algebra and applications of mathematics to augment an existing college geometry course. Under normal criteria these courses would hardly ever justify graduate credit, since they were designed to start where the in-service teachers were in those areas. However, it appeared that no viable alternative existed. Most of the mathematics teaching over the next several years would have to be done by teachers already in the classroom, with strong needs to extend or update their subject matter competence. Moreover, for those with the motivation and the ability to make up their deficiencies through summer course work or special night or weekend classes, the major traditional options usually had serious drawbacks. If school teachers took regular undergraduate mathematics courses with engineers, scientists, or liberal art students, relevance of the course content to their teaching areas would rarely be touched on; moreover, the credit earned would not help upgrade or renew their certification and thus not qualify them for badly needed pay increases. Nor would many of them fare well by electing to take research-oriented graduate level mathematics courses—even if they had the prerequisites, as few did. So more methods courses was the usual option, but this did nothing to improve their subject matter competence.

The role of NSF

The situation in mathematics in the early fifties was similar to that throughout the sciences. Among the many factors requiring teacher updating and curricular improvements even in the better schools were the electronic computer, nuclear sciences and space exploration. Thus when the National Science Foundation was formed, not only were college and high school science and mathematical faculties inadequately trained for traditional curricula, but also, most of the best existing faculty were not current in their knowledge of rapidly developing new scientific areas. Scientists of all disciplines were concerned about this situation, whether their primary functions were research, application, or teaching. So great was this concern in 1955 that Dr. Joseph Barker, then president of the scientific research honorary Sigma Xi, issued a call to action to all chapters to compete in proposing appropriate moves to overcome what he described as a major crisis in science education. More than 57 Sigma Xi chapters responded to this challenge. Of these, 8 proposals were ranked equally meritorious and published in the *American Scientist*, Vol. 44, 1956. These recommendations were concerned not only with the quality of science instruction but the parallel problem of recruiting adequate numbers of potential science graduates to serve in the age of computers, nuclear science and space exploration.

I chaired the committee from Alabama Polytechnic Institute (now Auburn University) whose report appeared among the winners. Our proposal was typical of efforts across the nation. We developed various ways of reaching students early in their high school years, including career opportunity programs prepared by honorary science organizations, preparation of newspaper articles, a series of educational television programs, and a speakers bureau to take campus scientists to civic clubs, schools' science clubs, and convocations. A major category of action was designed to alleviate the shortage of well trained science teachers by providing training opportunities like those mentioned above for mathematics throughout the science areas, and by influencing additional capable students to prepare for science teaching careers. Information about science teaching was transmitted to all freshmen education majors with high rankings on placement tests in mathematics and science. The report suggested cooperation at the national level with such organizations as Sigma Xi, AAAS, the Engineering and Scientific Manpower Commissions, the American Chemical Society, and the American Mathematical Society. The report concluded with a page citing promising starts around the country which might serve as models for expansion but, surprisingly enough, made no reference to the National Science Foundation which was then in its early years and had made no local impact to merit attention.

The following year, while I was on a sabbatical research grant at Maryland, Sputnik burst forth into the heavens. One consequence was a massive effort to beef up the NSF mission to improve science education and to recruit and train able scientists. In the summer of 1958 I was privileged to work on the NSF staff during the preparation of the science education budget which was at that time undergoing a ten fold increase!

Suddenly, it became possible to overcome many of the factors which inhibited the well thought out plans to improve instruction, develop curriculum, and attract high caliber students. For instance, based on the NSF budgets for 1957-1958 only three programs were funded throughout the country for secondary school students of high ability. As I recall, the prototype for mathematics in that trio at Florida State University (whose faculty I joined in the fall of 1958) attracted nearly 5000 applications for 40 positions even though very high selection standards were advertised. Soon there were scores of such programs each summer.

NSF expanded support for high school and college faculty institutes, as well as for curricular efforts such as the School Mathematics Study Group (SMSG) and the Committee on the Undergraduate Program in Mathematics (CUPM). Teacher institutes, short courses, and workshops contributed much to this effort. Proposals for funding were evaluated carefully to ensure that the content was predominately subject-matter oriented, although some method coverage (particularly for innovative curricula) was allowed. Many of the institutes awarded graduate credit which was appliable to advancing the teacher's status through degree awards or improved certification. These programs provided sufficient financial support to allow teachers to upgrade their qualifications without sacrificing needed summer income to supplement traditional low pay. As an illustration of the effectiveness of this method of attracting teachers to course work targeted for their deficiencies, a course I offered in applied mathematics for teachers which drew only 5 students as a regular summer offering a few years earlier drew 50 institute participants who demonstrably needed and benefited from it at an NSF institute in the summer of 1959.

It would take a large group of individuals to study the full range of the NSF programs for teacher training in mathematics over the span of Foundation activity, but I did have an interesting opportunity, in preparing a 1959 report commissioned by NSF, to study the proposals and the reports for all such programs conducted before 1958. The situation in the spring of 1959 with regard to in-service training of secondary school mathematics teachers was roughly as follows. The institutes funded for 1958 had appeared to leading mathematicians of the country to have a grossly inadequate percentage of openings for mathematics teachers, who constituted slightly more than half of the in-service teachers of science and mathematics in our country. (Social science teachers were then not covered under NSF proposals.) Thus the mathematics community made a concerted effort to generate more proposals for 1959 institutes. As a result, for 1959, 73 out of 150 of the single discipline institutes were exclusively in mathematics, and 125 of the 170 multiple field institutes included mathematics in their offerings. This meant a very rapid expansion of programs for mathematics teachers.

Unfortunately, too many of the institutes were single shot opportunities, not coordinated with degree programs or planned for a full upgrade of the participant teachers' needs. Numerous cross applications from potential participants occurred and no procedures for eliminating redundancy among repeat applicants from earlier summer programs had been established. In the summer of 1959 the mathematics teachers participating in summer institutes would be greater than the cumulative total over past institute history, and many from that increased total would be participants from previous years. As the program of teacher training institutes continued, it became obvious that planning was required to coordinate multi-summer participation for those who needed it. Very few of the local brochures implied a continuity of training possibility for their participants, although occasionally a brochure such as that of Rensselaer Polytechnic Institute would state, "The National Science Foundation sponsors a summer masters degree program in the natural sciences." An even more explicit claim by Union College was "The institute will offer masters degrees to all teachers who attend 3 institutes at Union and earn 30 hours credit." These were brave statements since no long range awards were authorized at that time.

Indeed, one major recommendation of my 1959 report to NSF was that multiple year funding for at least some mathematics institutes be provided to colleges developing a continuity of offerings which would permit a minimally trained teacher over a period of several summers to make up deficiencies in traditional areas and/or acquire familiarity with new topics slated for inclusion in the high school curriculum.

At the peak of science education activity the budget of NSF was almost equally divided between its two missions, research and instruction.

Some mathematics teachers participated in multi-discipline institutes offering only one course in mathematics, even though their needs were much greater in their own field of instruction. Thus, in 1958 the Mathematical Association of America's Commission on Summer Institutes stated, "In this time of curriculum modernization it appears important that secondary school teachers be provided the opportunity to give their full time study in a summer institute to mathematics." MAA also conducted short updating sessions for institute lecturers so they could benefit from the experience of pioneers from the early NSF prototype institutes.

At the peak of science education activity the budget of NSF was almost equally divided between its two missions, research and instruction. Prototype curricular development activities comparable to the Committee on the Undergraduate Program in Mathematics and the School Mathematics Study Group flourished throughout other science areas. The social sciences were added to the scope of NSF activity and interdisciplinary curricula involving natural and social sciences developed. Some of the new concepts in mathematics, such as set theory, logic, number systems, etc., began to appear in classrooms around the country. The so-called "new math" was still strange to many of those assigned to teach it and parents who had helped older children with their mathematics homework were frequently at a loss to understand or help with the new topics. Concurrently, many parents and law-makers developed similar adverse reactions to some of the social science theories advanced in new texts whose development was federally funded. Moreover, criticisms of other federally supported programs developed as some science enrollments caught up with demand, so continuation of existing enrollments were, or shortly would be, producing oversupplies of graduates in those fields. The net result of these criticisms was an over-reaction which eliminated major programs still needed to maintain quality in education and an adequate supply of graduates with current undergraduate and graduate degrees. During the eight years from 1968 to 1976 science education funding dropped from approximately half to less than 10% of the NSF budget. Although the overall budget was growing substantially to match inflation and increased demands for research, the undepreciated dollars for instruction declined to half of their peak amount, thus to much less than half on a depreciated dollar. Virtually all college faculty or pre-college programs such as summer and academic year institutes, short courses, and workshops were eliminated from the NSF budget. The various curricular studies were either phased out completely or retained on a highly limited basis for a few special case projects, as if important current developments in science and mathematics unknown to the in-service teachers would, automatically, be incorporated into their texts and teachings.

Throughout the 1970's proposals for college mathematics needs that were clearly recognized by leaders in the MAA and cooperation agencies could rarely be funded, no matter how highly evaluated by peers of the proposers. In 1976, out of 6 proposals which had been carefully planned to meet educational needs in mathematics by MAA officers and committees, not a single one was funded. In many ways, it began to appear as if the cycle of decline in mathematics education that occurred prior to NSF's establishment was being repeated. When these facts came to the attention of the Conference Board of Mathematical Sciences (CBMS) a Task Force on Science Education was named to enumerate priorities and set in motion efforts to restore adequate levels of funding for educational purposes in mathematics and the sciences. The Conference Board adopted the priority goals suggested by the Task force, and began a concerted effort to influence budget planning and appropriation levels which would make these goals possible. The action of CBMS was so similar to a parallel movement in the Council of Scientific Society Presidents (CSSP) that a close liaison has been maintained since 1978 between CBMS and CSSP to coordinate efforts to attain the priority objectives of the two groups. Henry Alder, who provided that liaison, has now become CSSP Chairman and named Richard D. Anderson as Chairman of the CSSP Education Committee. Thus mathematicians are playing a key role in the fight to restore quality level science education . . . again the Queen and Handmaiden of the Sciences.

As common efforts were being made to improve the quantity and quality of science education funding, a new threat to instruction in mathematics and science appeared. The administration bill to create a new cabinet level Office of Education as submitted to Congress in 1978 involved transfer of almost all of NSF's decimated education programs to the new office. This would have completed the cycle of downgrading science education to the status it had when the Foundation was created; the fate of science and

mathematics education would again be determined by experts in methods and evaluation rather than by those with current knowledge of subject matter. Through the action of the House Committee on Science and Technology, the bill did not clear in 1978. The NSF deletion was omitted from the House version of the bill and no action to compromise was taken. Under heavy administration pressure, and in spite of much testimony before cognizant committees in both Houses by scientists and mathematicians, versions of the 1979 House and Senate bills contained provisions to transfer nearly half of NSF's programs to the new Office of Education. The half chosen in each chamber overlapped very little and renewed contacts with the compromise committee were successful in the final bill to limit the transfer to only a small fraction of the existing programs, those for elementary teachers and certain special groups.

Computers and the mathematical sciences

No single development has had as much impact on mathematics in the latter half of this century as the electronic computer. So pervasive has been its effect that academic departments of computer science or data processing have sprung up in departments of electrical engineering, in schools of library and information science, in business schools and technical schools, as well as in those mathematics departments which provided proper recognition and encouragement. The failure of many mathematics departments to welcome computer science as a new component was a major factor in the impetus towards establishment of separate departments of computer science. Although none of my doctoral studies in mathematics was oriented toward numerical methods or computation in any manner, I have been concerned since the early 1950's with the potential impact the computer should have on the mathematics profession and the teaching of mathematics at all levels. This interest led naturally to various computer experiences in the late 50's, which themselves led to an appointment throughout the 60's as the first director of the Computing Center of the Florida State University. The interactions between mathematics and computing at Florida State in the 20 years since I started my tour of duty in the University Computing Center reflect closely national trends during that time, and illustrate well the role that the NSF played in those changing times.

After a year and a half of intermittent discussion the department of mathematics at Florida State finally approved, in 1962, its first credit course in computing—a one hour course in Fortran programming which included an approved term project to produce and document a program to perform a significant task, preferably from the student's major field of study. A high percentage of these term projects were in mathematics or applications of mathematics. Books of these sample programs were compiled by the instructors and indexed by major fields in a local project called TACT (Teaching Aids for Computing Techniques). A campus-wide committee was formed to approach non-computer-oriented faculty members with TACT, offering to provide a human aide to go with the TACT material to discuss computer solutions for problems of the type studied in an existing course which they taught. This outreach to faculty members through the work of their students had limited success, depending as it did on volunteer help, but efforts to fund a more substantial implementation through NSF or other sources failed. At that time NSF policies specified that curriculum development projects would be funded only for organizations of national scope where high qualifications and prestige of the proposers and of the participating faculty would enhance the acceptance and dissemination of the project's output. Meanwhile our TACT modules accumulated with little use, even though many of them, if properly modified, would be suitable for classroom use in many undergraduate mathematics courses. Furthermore the successive reports of CUPM were urging curricular modifications in those same courses which would reflect the impact of the computer.

Soon Florida State University joined sixteen major universities to plan a Center for Research in College Instruction in Science and Mathematics (CRICISAM) for which our campus was later selected as host. The site selection committee was quite interested in our local computer efforts and in part based their decision to establish headquarters at FSU on that interest. Before the details of the CRICISAM establishment and location were complete and publicly announced, the National Program Committee for ACM and the Southeastern Region of SIAM organized a national symposium to discuss the potential role of computers in the instruction of undergraduate mathematics. About 50 mathematics department heads or their delegated representatives attended, along with an equal number of other interested individuals. William Atchison discussed the core mathematics courses planned for the emerging curriculum in Computer Science which was being developed by the ACM sponsored, NSF funded, Curriculum Committee on Computer Science (C^3S). He quoted from the preliminary recommendations from this Committee, which ultimately produced Curriculum '68. He also cited 2 CUPM reports: the first for engineers and physicists, the second for work in computing. (At that time the FSU Mathematics Department was adding the necessary courses to allow its mathematics majors to obtain a degree meeting fully the recommendations of that latter CUPM recommendation and planning, with the Computing Center, appointment of faculty who could develop computer science offerings in keeping with these curricular goals.) The other speakers at the symposium looked more to the problem of students majoring in mathematics who would benefit from computer-oriented offerings in appropriate courses. The calculus sequence was frequently cited as an important place to include computer related materials and computer implemented methods; linear algebra, differential equations and other courses were also suggested as prime targets for modernization.

The first public announcement of the formation of CRICISAM was made at that ACM-SIAM conference. Its first major effort was the production of a prototype computer-oriented calculus text. The NSF funded CRICISAM calculus texts which resulted from this project made a major impact on the teaching of calculus throughout the country. The book, whose authors were Warren Stenberg and Robert Walker, was used for quite a number of classes on more than 100 campuses, many of which adopted the text over a period of several years until alternate computer-oriented commercial texts were available. This dissemination of NSF sponsored instructional material was greatly aided by two NSF funded College Faculty Summer Institutes at FSU in 1969 and 1970. These institutes took

---

Things were looking pretty good for mathematics education . . . as we moved into the 70's, but the bottom was just beginning to fall out.

---

established mathematics faculty with no significant background in computer programming or computer use and gave them enough training in programming and in the concepts of the CRICISAM calculus so they could return to their own campuses and make an offering of computer-oriented calculus courses in the following academic year. In fact, they and their department heads were committed to such action before they could be accepted for institute participation. Some institute funds were used to conduct a spring 1970 conference at which the participants from the previous year could exchange information on problems they had encountered and advice they could give to the next participants as to how the computer-oriented calculus course should be taught for greater effectiveness.

Things were looking pretty good for mathematics education on the FSU campus and throughout the country as we moved into the 70's, but the bottom was just beginning to fall out. CRICISAM had obtained a small planning grant to work up a prototype computer-oriented book in ordinary differential equations. A strong National Advisory Committee was lined up and the project seemed clearly consistent with national goals for curricular improvement in the sciences. Nevertheless, the grant was denied and very soon the NSF program for curricular development was virtually eliminated. CRICISAM quickly folded from fiscal malnutrition. Although there was still much demand on the part of college teachers to participate, a proposal for another CRICISAM Calculus Institute was not funded and soon thereafter college faculty institutes were eliminated altogether.

Computer related mathematics education at FSU did get some benefit for 1976-79 from a new NSF program established under congressional pressure, the Comprehensive Assistance to Undergraduate Science Education Program (CAUSE). This program made it possible for us to develop and test computer-oriented modules for a major segment of the undergraduate mathematics curriculum. We prepared printed handouts, console demonstrations, interactive programs, or sequences of computer assisted instruction (CAI) programs for existing courses, including elementary matrices; differential, integral and numerical calculus; ordinary and partial differential equations; modelling theory; complex variables; and the structures courses in computer science. By a 1980 decision, computer science is now an equal partner with pure and applied mathematics in a single department which has voted to change it's name to the Department of Mathematics and Computer Science. So in the past 20 years the department has moved from no courses in computer science to a dynamic state where majors in computer science now outnumber their pure or applied counterparts at both the undergraduate and graduate levels. Had it not been for the considerable NSF support, the department could not have made such a move and certainly would not have voted to do so. With budgetary constraints holding faculty positions level, and actually cutting them at times, this growth was possible only because several able tenured faculty members took the trouble to learn enough about computing to take over some of the course loads in computer science. Other departments, without NSF support, may have great difficulty adapting to the new challenges of computing and computer science.

Policies for the 80's

It appears that some of the forces contributing to the decline of science education funding have come from the constraints placed upon the administration and staff of the National Science Foundation by the Office of Management and Budget and other facets of external administrative control. When a budget is submitted to Congress, it is the President's budget; agencies headed by political appointees are obligated to support it, even if they realize that it fails to provide for clearly established needs of the country which should be of high priority under any qualified appraisal. Science education programs are often eliminated from preliminary budgets, sometimes in several successive submissions, with such justifications as, "It is not administration policy to subsidize higher education." Of course, if this were true, it would have wiped out most activity in support of college level education in all agencies, a stand which few would try to defend. Another frequently offered reason for rejecting some proposals for development of innovative curricular material is that it would be better if such

materials were produced by commercial publishers to insure their market-ability. Proposals from professional organizations that needed overhead funds to carry on additional projects took a real beating in the reviews as being excessively expensive even when strongly supported by competent advisory committees. Yet innovative instructional material, sensitive to both the needs of the time and to the state of the art in science and technology, usually needs the contribution and judgment of experts as well as the prestige of appropriate, professional organizations. Even more impor-tant, when such materials are produced, are mechanisms for acquainting teachers in their use. In eliminating essentially all curricular development projects of national stature as well as most institutes, short courses, and workshops for updating instructors, the NSF demolished both phases of the successful symbiotic processes for maintaining currency in science and mathematics instruction. Commercial publishers on their own devices will do very little to carry on work required to maintain currency of subject matter.

For the first few years of this downward cycle of science education funding, research scientists did not speak up as they should to protect the important educational aspect of the nation's scientific effort. They were pleased to see growth for research in their own areas and perhaps didn't realize immediately what was happening to threaten the nation's supply of well-trained young scientists. Perhaps also, the growth of the U.S. Office of Education and its threat to take over the science education programs at NSF caused the NSF administrators to minimize budget requests in the threatened areas.

> *In eliminating essentially all curricular development projects of national stature . . . the NSF demolished both phases of the successful symbiotic processes for maintaining currency in science and mathematics instruction.*

Fortunately, the situation seems now to have changed, both among the scientific leaders and the NSF staff. Among the organizations which constitute the Council of Scientific Society Presidents and the Conference Board of Mathematical Sciences, research-oriented groups far outweigh instructional organizations. Yet, both the Council and the Conference Board have unanimously supported strong resolutions calling for major increases in science education funding for curricular development and faculty training programs. They have elected into important offices individ-uals who have played key roles in identifying those educational needs and will continue to fight for them until more adequate funding and program authorizations have come into being. Furthermore, in early 1980 the education staff of NSF prepared for the National Science Board a strongly worded and well-documented appeal for support to restore activities in curricular development and teacher training to much more appropriate levels. Discussion of these actions in early 1980 suggest that key NSF or National Science Board leaders might have done more to prevent the present low ebb if voices of encouragement had been heard earlier from leading scientists and from key members of the administration and of the legislative bodies. Now that the research scientists are speaking out, through their professional organizations, their views should gradually percolate up through the Office of the Presidential Science Advisor to the appropriate Congressional leaders and to the Office of Management and Budget. This advocacy may initiate a new upward cycle leading to an effective level of support for scientific education. If all those concerned with research, development and scientific education get behind current efforts to bolster our sagging support for science education, science can once again play a critical role in our national effort to improve the quality of life and to solve our problems at home and abroad.

The above remarks, prepared nearly a year ago, ended on a hopeful note which, at the present, seems optimistic. An extensive report commissioned by outgoing President Carter in 1980, released almost on the eve of his election defeat, clearly showed the great national need to bolster sagging educational support in science, engineering, and the mathematical sciences upon which the others so heavily depend. Yet the 1982 budget, which just went forth to Congress, lowers proposed Science Education funds to a 30 year bottom of 7.5% of the NSF Budget. The *Science* Editorial for 23 January 1981, expressing alarm at this abysmal retreat in the face of the enemy, technical inadequacy, closed with this statement: "The Reagan Administration has the opportunity, without compromising prudent economic policies, to reorder priorities and set a positive course toward rebuilding America's excellence in science and engineering education." Let us hope that this opportunity will be taken.

Issues of Equality

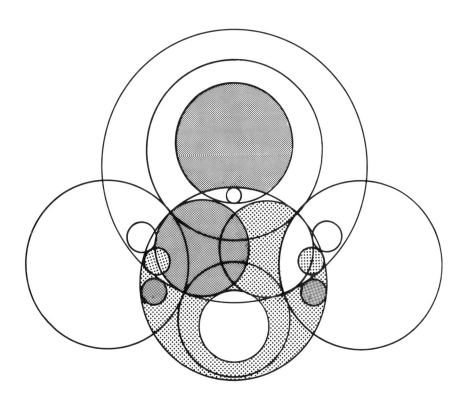

The Real Energy Crisis

Eileen L. Poiani

She stood at the podium wearing a pastel dress stiffened by yards of crinolines, hands shaking a little, internal butterflies fluttering. The audience of about 300 of her classmates and 1,000 parents and teachers was hushed and awaiting the beginning of the annual ninth grade promotion exercises. She took a deep breath and began her speech:

> The launching of the first Sputnik by the Russians and the sudden awareness of Soviet superiority in one area of science has had the effect of a stunning surprise on the American public . . . . Our schools will be called upon to supply an increasing number of persons with a wide range of skills, many in new fields resulting from the growth of technology. The welfare and security of our people as a whole may well depend on the extent to which we are able to educate each young man and woman to his or her full capacity . . . .

That program took place more than 23 years ago. The message was clearly in reaction to the Sputnik gauntlet laid before the United States on October 4, 1957. And she was I.

As I gathered, tested, and rejected ideas with which to begin this essay, I kept returning to that moment in the school gymnasium in 1958. Why? Because it represents for me the vision of many issues which would engage me in the years which followed, issues which I would like to share with others—parents, teachers, school counselors, mathematicians, public decision-makers, and interested readers.

Eileen L. Poiani is Professor of Mathematics and Assistant to the President at Saint Peter's College in Jersey City, New Jersey. She received her bachelor's degree from Douglass College in 1965 and her Ph.D. in mathematics from Rutgers University in 1971. A member of the Saint Peter's College faculty since 1967, she is now also responsible for long-range planning and institutional research. Since the program began in 1975 she has served as the National Director of WAM: Women and Mathematics, a lectureship program sponsored by the Mathematical Association of America under a grant from IBM. She is a Councillor of Pi Mu Epsilon, the National Honorary Mathematics Society, and a former Governor of the New Jersey Section of the MAA.

To be sure, the thrust of Sputnik did lead to what Neil Armstrong in 1969 called "one small step for man, one giant leap for mankind," as he stepped from the *Eagle* of Apollo II onto the lunar surface. Within a decade American technology had advanced meteorically. Concurrently, however, few children were being educated to their capacities, and faults in the American dream began to quake: fundamental institutions were challenged, college campuses exploded, civil rights were demanded, and Betty Freidan's *The Feminine Mystique* emerged, to name a few. What happened in those intervening years would take years to explore, and you did not choose this book for that purpose.

But what was and what remains "relevant" (remember that word?) is the fact that mathematics—from arithmetic to topology, from pure to applied —undergirds the technology of space and of earth. While people in all walks of life feel comfortable operating the products of that technology— washing machines, cameras, TV computer games, tape recorders, CB radios, microwave ovens, calculators—they are turned off by the very mathematics that underlies them. Ironically, one out of five Americans today is said to be functionally illiterate! They lack the basic skills required to function effectively in daily life—skills like checking the arithmetic on a bill of sale, reading the classified ads, and understanding a train schedule. Recent studies here and abroad verify that literacy in mathematics (or numeracy as the British call it) from arithmetic through geometry and algebra, together with public understanding of its usefulness, is sorely inadequate. Even outcries against the metric system reflect the public desire to avoid anything mathematical.

People do avoid mathematics. They stereotype mathematicians as introverted, cold, odd, genius-type, dull, unattractive. People associate these unpleasant characteristics with mathematics itself and hence want to avoid it. At social gatherings when asked "What do you do?," most mathematicians have had their replies met with blank stares or proud true confessions: "Ugh! I hate math," "Math was my worst subject," "You must be a genius." Backed into a defensive posture, the mathematician either proceeds to extol mathematics' virtues or retreats to a less hostile corner.

Many jokes, cartoons, and one-liners malign the liking of mathematics— even mathematicians chuckle at them! I cannot forget this quotable one from a popular women's magazine: a ten-year old asked her four-year old sister what she wanted to be for Halloween, adding it should be "something scary." The child replied, "How about a math teacher?" Unfortunately, the real joke may be on society, since mathematics will be needed in increasing proportions as the 21st century approaches.

It is a fact of life that solving mathematics problems can be frustrating. Some people can tolerate more frustration than others. Frustration can trigger fear, anger, boredom, and a host of emotions that lead to avoidance of their cause. A possible consequence of these feelings is the phenomenon of "math anxiety" which has been popularized recently through the writings of Sheila Tobias and Stanley Kogelman; an earlier manifestation of the same syndrome was dubbed "mathophobia" by Jerrold Zacharias of the Massachusetts Institute of Technology around the time Sputnik was launched. Symptoms range from mild aversion to sweaty palms, heart palpitations, and even nausea. While a mild case of the jitters can help the adrenalin and creative juices flow, excessive anxiety can throw up serious mental blocks. Estimates suggest that about half of the population may be victims, two-thirds being women! There are many causes:

— *Negative childhood experiences.* For example, if an elementary school teacher rewards the child who solves a mathematics problem quickest, the slower worker may feel embarrassed or discouraged.
— *Punishment.* Requiring a child to stand at a blackboard until a problem is solved (or the child dissolves in tears), or assigning 30 mathematics problems for talking out of turn, causes children to associate mathematics with punishment. This creates anxiety by association.
— *Excessive absences from school.* The child who leaves class studying division may return to percents. Confusion and fear result. One of my college students said that she was absent when her class studied addition with 9's and to this day has trouble with it.
— *Contagion.* Anxiety is contagious. Parents have it. Teachers have it. Counselors have it. Kids catch it.
— *Choice.* It is perfectly normal for a person to feel tense when making a decision. Just remember what it's like to choose options on a new car. Pity the mathematics student, head swimming with formulas and procedures. Enter a mathematics problem; even worse, a *word* problem. (Mathematics lovers and haters alike shudder at the mention of that word.) Choosing the right procedure from all those learned is a nontrivial, anxiety-producing task. At least a half-dozen thought processes are involved in just trying to factor an expression like $x^2 - x + 6$.

The nature of mathematics itself contributes to public avoidance and to its isolation from general society. One problem is that mathematics requires solitude. Although you can and should talk about mathematics, when it comes to the nitty-gritty, you have to sit down and do it alone. Young Americans live in a gregarious society; the loner is the exception and few want that stigma. Another problem is its technical language. Have you ever been in a room where everyone is speaking a language foreign to you? You feel left out and somewhat intimidated. People feel the same way when they hear unfamiliar mathematical jargon.

Teachers and advisors, however unintentionally, often undermine natural enthusiasm for mathematics and inhibit development of mathematical talent. Some teachers demand that students solve mathematics problems *only* by the method taught in class. This process can thwart initiative, compound frustration, and promote avoidance.

In their zeal to recruit from a shrinking applicant pool, college admissions officers are heard to remark, "You don't need mathematics to get into College X." Sad but true, this appeal can stifle the student's interest in taking more high school mathematics or accede to the student's desire to avoid mathematics and thereby limit his or her career horizons. School counselors frequently discount the fact that mathematics is really needed for entry or advancement in nearly every occupation —whether or not it requires a college background. Like it or not, mathematics opens career doors, so it's downright practical to be prepared.

Women and minorities have traditionally composed the largest group left out of the mathematical mainstream. Cultural stereotyping of the mathematician as masculine has conditioned women to avoid mathematics, and society to discourage women from mathematics-related pursuits. As Plato points out in *The Dialogues*, "the state, instead of being a whole, is reduced to a half" when men and women fail to follow the same pursuits. This is indeed the real energy crisis.

Like it or not, mathematics opens career doors, so it's down-right practical to be prepared.

Since the proverbial apple does not fall far from the tree, generation after generation of women have avoided mathematics. Surely some men are anxious about mathematics, but few admit it. Fewer still avoid it. Society expects men to enter professional careers or technical fields which need mathematics, so teachers and counselors steer them into the mathematics curriculum and the "Pygmalion effect" of students performing in response to teacher expectation takes over.

Usually avoiding all but the minimum high school mathematics requirement, young women with aptitude have long been inadequately prepared to take college calculus. Today, nearly 300 years after the discovery of the calculus, sociologist Lucy Sells has identified mathematics as the "critical filter" in career access for both college and non-college bound students. Perhaps the most frequently cited statistics—already classic—that support this claim are those which Lucy Sells uncovered: 57% of the men in the freshman class at the University of California at Berkeley in 1972 had taken four years of high school mathematics compared with a mere 8% of entering women. At that time, all major tracks at Berkeley required four years of mathematics, except, as you might suspect, librarianship, education, social sciences and humanities. So 92% of the women had to settle for these traditionally female, lower paying fields.

Mathematics plays the critical filter role not only at the college level but also in the work force. According to the *Occupational Outlook Handbook*, the U.S. Bureau of Labor Statistics forecasts that the vast majority of the 46 million job openings through 1985 will require fewer than four years of college training. Major growth areas include computers, transportation, health services, secretaries, bookkeepers, cashiers, mechanics, and police officers. These areas require good basic arithmetic skills and demand an ability to handle data, understand statistics, read computer printouts and graphs. Performance on employment tests and advancement in these fields may well depend on a person's mastery of algebra and geometry. In fact, in the November 1977 issue of *Today's Secretary*, it was pointed out that "if you want to advance in a secretarial career, math may be one of the most important hurdles to overcome to be successful. Today, the highest paid secretaries are in the technical fields." The ability to handle budgets for the boss is an asset when seeking advancement to be the boss.

From my own personal visits to high schools, I have found enormous differences in the attitudes of school personnel towards the mathematical persistence of women. Remarks from a counselor in a sparkling suburban secondary school were disturbing: "If more women would just stay home and raise families, this world would be a better place." Contrarily, there are many champions of the cause in the schools. Their zeal is crucial for overcoming the overt and subtle efforts that dissuade young women from the mathematical mainstream.

The "fear of success" syndrome and the career/family conflict continue to surface in questions asked by high school women. They have the feeling that the young men just won't date them if they are whiz kids in mathematics or science. Sad but true, these fears have been validated in the past. In 1976, Edith Luchins' survey showed (see [5]) that nearly half of her 350 respondents (all women mathematicians) had been discouraged by family and school personnel from studying mathematics, a frequent reason being that "the boys wouldn't like them or date them." Fortunately, the next generation seems more open to the new roles of women.

A supportive environment seems to prevent premature withdrawal from mathematics according to a study by Patricia Casserly of the Educational Testing Service (see [2]). In viewing secondary schools where women persist and excel, she found the following factors account for success: role models, peer tutoring, parental and teacher awareness of mathematics' usefulness, and avoidance of the "tear trap"—treating a frustrated student with understanding but not excusing her from doing or taking mathematics.

How are people reacting to all the fuss about women and mathematics? The communications media—a former math avoider—is now calling attention to these issues through newspaper and magazine articles, radio and television interviews. People are cancelling out their fears and honing their skills through clinics and workshops on several college campuses, and through private counseling programs.

The International Business Machines Corporation (IBM) has had a strong and enduring reaction. IBM representatives hosted a reception for top scorers on the U.S.A. Mathematical Olympiad and noticed that no women were among the winners. The U.S.A. Olympiad is an annual contest begun in 1972 for invited high school students who excelled in previous mathematics competitions. From the high achievers in the U.S.A. Olympiad, a team is chosen to compete in the summer International Olympiad. American teams have entered the international contest since 1974, but no women have yet been on the team. (Two did participate for the first time in the training sessions in 1978 and each year since then.)

Mathematics plays the critical filter role not only at the college level, but also in the work force.

What Sputnik failed to do for women, IBM has helped to do. Absence from the Olympiad symbolizes their absence from fields requiring sound mathematical preparation—traditional fields such as engineering and physics, as well as newer areas like accounting, business management, medicine, computer science and the social sciences. So in 1975 IBM agreed to support a secondary school lectureship program of the Mathematical Association of America to encourage young women to take more mathematics and keep career doors open. Thus, WAM (Women and Mathematics) was created.

In order to provide flexibility in meeting the needs of each request, the WAM program functions on a regional model. There are presently nine active regions centered in Boston, Connecticut, New York/New Jersey, Chicago, South Florida, San Francisco Bay, Southern California, Oregon, and Seattle. Other corporations have also joined forces with IBM to contribute to the support of WAM, and if additional funding becomes available, plans for expansion to new regions will be implemented.

WAM reaches ninth and tenth grade audiences in visits of half to one full day at no cost to the school. Speakers serving as role models are drawn from all walks of life. Their occupations represent the worlds of business, medicine, academe, research, industry, government, and skilled trades. In formal and informal talks with students, teachers, and counselors, they stress why mathematics is necessary for both college and non-college bound students regardless of sex.

Presentations are also arranged for professional societies, elementary school teachers, civic organizations, parents groups, and legislative leaders. Since guidance counselors exert such a profound influence on students, a special one day conference named MODE (Math Opens Doors Everywhere) was designed for them. Its goals are to disseminate materials on the usefulness of mathematics, to provide a forum to exchange ideas, to discuss how to combat the masculine image of mathematics, and to explore how to overcome math anxiety. MODE conferences have been held in several WAM regions.

By conservative estimates at the time of this writing, WAM has reached more than 700 schools, 56,000 students, and 6,800 parents, teachers, and counselors, since it began in Fall, 1975. Students, hosts, and speakers evaluate each visit. A detailed analysis of WAM effects has also been the subject of a doctoral dissertation by WAM Coordinator, Carole Lacampagne. The outcomes have been gratifying, and some high schools have attributed to WAM an increase in subsequent female enrollment in advanced mathematics courses.

Here is the story of a young college woman whom I taught in calculus. Her struggle is what WAM hopes to prevent:

> Math was not required in high school so I took only one year. I was first generation American and crossed a language barrier in school. Even if someone had reached my parents, they could not help me. I worked a few years as a secretary keeping stock inventories on securities traded in foreign markets. I learned the hard way with the help of very patient people. But I learned mechanics and I still have trouble seeing math's essence . . . . Only a person in the field can convince a high school student about the realities of a job. My "cheers" to your program.

This story has a happy ending (really a beginning!). She is now a medical student.

(A new Blacks and Mathematics (BAM) program, modelled after WAM, is sponsored by the Mathematical Association of America under a grant from the Exxon Corporation. Operating in Atlanta, Connecticut, Detroit, Houston, and Washington, D.C., its primary goal is to encourage black students to enroll in college-preparatory mathematics.)

The trend of men outnumbering women by 2:1 (sometimes as high as 3:1) in junior and senior level mathematics classes is beginning to reverse. A national study by the Education Commission of the States gives cause for optimism. Based on a survey of 1500 thirteen-year-olds and 1800 high school seniors, results show that sex differences in participation in mathematics courses have diminished noticeably in recent years, with 41% of the men and 37% of the women surveyed having had four years of mathematics. In terms of achievement, thirteen-year-old females began high school with at least the same mathematical abilities as males and were even better in computation and spatial relations (a controversial area). By senior year, however, the males surpassed them in problem-solving and were on a par in the other areas.

National data shows that the sex difference continues in scores on the College Board Mathematics Scholastic Aptitude Test. The difference still hovers around 50 points, although the gap is narrowing slightly each year with 1980 senior women scoring 48 points less, on the average, than their male counterparts. Recent evidence suggests that societal conditioning, opting out of mathematics, and even the shape of childhood toys contribute to the poorer performance of females rather than innately inferior ability. Further research on these issues is continuing.

Progress is further evident in a 1980 Census Bureau report comparing college majors of 1978 with those a dozen years earlier. Four to six times as many women were majoring in the traditionally male-dominated fields of business and engineering in 1978 than in 1966. The shift is slow but sure.

A series of panels and lectures at the 1980 International Congress on Mathematical Education probed the progress and problems related to the status of women in mathematics. Just a few short years before, no such issues appeared on programs of meetings of the mathematical community.

If WAM continues to help smash the erroneous stereotype that women just don't belong in mathematics, then the need for this program and its counterparts will be overcome. The real goal of WAM is to put itself out of business. But WAM cannot do it alone. WAM needs help both from those who love mathematics and from those who have left it. Here are some strategies:

— *For Parents.* "Protect your children's future. Have them take algebra and geometry in high school, whether college bound or not." So says a California public service announcement!

— *For Teachers.* Humanize mathematics. Sit back a moment and think of the name of a famous mathematician. Other than Einstein, few of these names are household words. I would hazard a guess that yours was on the side of Adam rather than of Adam's rib. So let the world know that "real people" of both genders with all the normal anxieties, frustrations, and joys create mathematics. Study the *Who's Who* and begin dropping names like Hypatia, Maria Agnesi, Sonya Kovalevsky, and Emmy Noether.

— *For Counselors.* Emphasize how and why mathematics opens career doors everywhere, and math avoidance closes them!

— *For Mathematicians.* Continue to erase the inequities within your own discipline, the barriers between pure and applied, and the controversy between the "old" and the "new."

— *For Business and Industrial Leaders.* Reappraise the requirements of your employees' jobs in terms of mathematical and logical thinking skills. Could deficiencies here be decreasing your productivity and preventing upward mobility of otherwise qualified employees? Would it be in the best interest of your company and your employees to establish temporary remedial mathematics programs? After all, even thinking metric is easier when thinking math is not anxiety-producing.

— *For Decision-Makers.* Assess carefully the merits of state and local pressures to shape mathematics education. The importance of quality pedagogy and curriculum design in mathematics cannot be overestimated.

— *For the Public.* Dispel unfounded mathematics myths and stereotypes.

Promote the fact that mathematics itself is blind to the color, neutral to the sex, and oblivious to the name or creed of the person doing the mathematics.

There is no denying that Sputnik set off a technological chain reaction that only now is reaching human proportions. Years after that speech in 1958 I became pointedly aware of the apparent anomaly of being a woman and having such technological visions. It was a stunning awakening. But today the twain does seem to be meeting and, step by step, mathematics—the true equalizer—is gaining acceptance as the critical ingredient in educating each young man and woman to his or her full capacity.

References

[1] Grace M. Burton. "Regardless of Sex." *Mathematics Teacher* 72 (1979) 261-270.
[2] Patricia Lund Casserly. "Helping Able Young Women Take Math and Science Seriously in School." *New Voices in Counseling the Gifted.* Ed. by Nicholas Colangelo and Ronald Zaffrann. Kendall/Hunt, 1979.
[3] John Ernest. "Mathematics and Sex." *American Mathematical Monthly* 83 (1976) 595-614.
[4] Stanley Kogelman and Joseph Warren. *Mind Over Math.* Dial Press, New York, 1978.
[5] Edith H. Luchins. "Sex Differences in Mathematics: How Not to Deal With Them." *American Mathematical Monthly* 86 (1979) 161-168.
[6] Lynn M. Osen. *Women in Mathematics.* MIT Press, Cambridge, Mass., 1974.
[7] Eileen L. Poiani. "Analysis of MODE: Math Opens Doors Everywhere." *American Mathematical Monthly* 87 (1980) 462-465.
[8] Sheila Tobias. *Overcoming Math Anxiety.* Houghton-Mifflin, Boston, 1980.

Women and Mathematics

Alice T. Schafer

In any history of mathematics one is hard pressed to find mention of women mathematicians. One wonders if there were any and, if any, why so few? To answer this question, it is perhaps necessary only to consider the lives of some of the few there were. We do this in the first part of the paper, when we look at the lives of six women mathematicians from six different countries, the first born in 1718, the latter dying in 1935.

There are, of course, related questions. If there were only a few women mathematicians before the twentieth century, how many have there been in the twentieth century? What has been done, and what is being done, to increase their number? What is the prognosis for the 1980's? These questions will be answered in the second half of the paper when we consider the role of women in the mathematical community in the United States during the twentieth century.

There is not space in this essay to go into any of the details of the mathematical work of the six women mathematicians whom we shall treat. Instead the focus will be on the barriers which stood in their way, and the struggles which they underwent in attempting to follow their desire to be mathematicians.

Alice T. Schafer recently retired as Helen Day Gould Professor of Mathematics from Wellesley College. She received a B.A. degree from the University of Richmond in 1936 and a Ph.D. degree in mathematics from the University of Chicago in 1942. She was awarded an honorary D.SC. degree from the University of Richmond in 1964. Prior to joining the Wellesley faculty, she taught at Connecticut College, Swarthmore College, the University of Michigan and several other institutions. She was the second president of the Association for Women in Mathematics (1973-75), and from 1976-78 was Director of the Wellesley College-Wesleyan University Mathematics Project to Combat Mathematics Avoidance in College Students. She is currently a member of the AMS-MAA-NCTM-SIAM Committee on Women in Mathematics of which she was the chair from 1975 to 1981.

Six women mathematicians, 1718-1935

Marie Agnesi (1718-1799), born in Milan, lived at a time when it was acceptable for women in Italy to be educated, contrary to the customs in other European countries. She was the eldest of 21 children born to a literate family. She was a very precocious child, and her mother and father, the latter a professor of mathematics, encouraged her. At the age of nine, at one of the gatherings of intellectuals in her home, she delivered in Latin a discourse defending higher education for women.

Her best known work, *Treatise on Analysis for the Use of Italian Youth*, dedicated to the Empress Maria Theresa, was published in two volumes in 1748. She had been tutoring her younger brothers and sisters, and this work was written for them, but others soon discovered it and it was translated into many languages and used as a textbook. She had collected much of the then known work on plane curves, the calculus, and differential equations for these volumes. Because she had a mastery of many languages, including Latin, Greek, and Hebrew, she had been able to read the papers in which research on these subjects was reported. It was in the *Treatise* that she discussed the cubic curve $(x^2 + a^2)y = a^3$ which became known as the Witch of Agnesi, due to a misunderstanding. Today many students who study analytical geometry know at least the name Agnesi.

Her work was sufficiently well known in France that a committee of the French Academy of Sciences was appointed to assess it. One member wrote to her, as quoted by Mary R. Beard in *On Understanding Women*:

> I do not know of any work of this kind that is clearer, more methodical or more comprehensive. . . . There is none in mathematical sciences. I admire particularly the art with which you bring under uniform methods the diverse conclusions scattered among the works of geometers reached by methods entirely different.

Although members of the French Academy thought highly of her work, had they wanted to invite her to join the Academy they could not have done so, for membership in the Academy was denied to women. There is a question among historians as to whether Agnesi ever held a chair at the University of Bologna, but it is clear that after her father's death in 1752 she devoted herself completely to religion and to aiding the poor.

While Agnesi was still alive, another woman, Sophie Germain (1776-1831) destined also to be a mathematician, was born in Paris. She was not as fortunate as Agnesi in her choice of parents. They were prosperous, and could have allowed her to study whatever field she wished; they did indeed encourage her in intellectual pursuits until she chose mathematics. That was too much. When they discovered that she was secretly studying mathematics in her room at night, they took away her candles, her fire, and her clothes, leaving her only her bedcovers. She managed to secrete candles

in her room and, after the other members of the family were asleep, she continued her study of mathematics. In the meantime she taught herself Latin so that she could read some of the mathematics books she had obtained. After finding her cold and asleep at her desk many times, her family gave in and allowed her to study mathematics.

It was just at this time that the École Polytechnique was founded, and Germain looked forward to studying mathematics there only to learn that women were excluded from attending the Polytechnique. However, lecture notes were available to all who asked, and Germain obtained these. Students were also allowed to submit written observations, which Germain did under an assumed name, M. LeBlanc. In this way communication between her and the well-known mathematicians of the day was started. In fact, J.L. Lagrange, who was at the École Polytechnique, was so impressed with "his" work that he insisted on meeting "him." When Lagrange discovered that M. LeBlanc was a woman his respect for her work continued. Through Lagrange, Germain got to know all the French scientists of the day, and her home soon became a center for meetings of some of the most distinguished of the group.

In 1809 Napoleon ordered the Academy of Sciences to offer a prize for a solution to the problem of finding a mathematical theory for elastic surfaces and comparing it with experimental data. Twice, anonymously, Germain submitted solutions; the first time her entry was the only one submitted. Each time a mistake was found and no prize was awarded. The third time she submitted an entry using her real name and won, despite a lack of rigor due to lack of formal training. Her equation for elastic laminae is still the fundamental equation of the theory and is now known as Germain's equation. She had won a prestigious prize without having published a single paper.

By now she was known under her own name and had begun to publish the results of her research, which were in many different fields. For example, she solved a special case of Fermat's last theorem, that a solution to the equation $x^n + y^n = z^n$ is impossible in positive integers for any integer n greater than 2. Her work on this problem was used by L.E. Dickson in 1908 to prove Fermat's last theorem for every odd prime n less than 1700. In 1910 E. Dubois named a special type of prime number as a *sophien*, thus ensuring that Sophie Germain's name lives on in the theory of numbers, too. She also worked on curvature of surfaces, and was even a philosopher; after her death, a nephew collected her writings in philosophy and published them under the title *Considerations générale sur l'état des sciences et des lettres aux différentes époques de leur culture*.

Germain never held a post at an academic institution. The story of her correspondence with the famous mathematician C.F. Gauss, whom she never met, is well known. A mathematician friend recommended to her Gauss's *Disquisitiones Arithmeticae* (published in 1801), although at that time not many mathematicians had been able to penetrate deeply into the work. Germain was greatly interested in it and in 1804 wrote her first letter to Gauss, including some problems in number theory on which she was working at the time. She again used the assumed name of M. LeBlanc; later, by accident, Gauss learned that LeBlanc was a woman, but, to his credit, their correspondence continued. Over a four year period they exchanged many letters concerning mathematical questions. Gauss recommended to the faculty of the University of Göttingen that she be awarded an honorary doctor's degree, but unfortunately, she died of cancer, in 1831, before it could be conferred.

Four years after the birth of Germain in France, Mary Fairfax Somerville (1780-1872) was born in Scotland. She grew up in a small seacoast village and received no formal education until after she was ten years old, when her father sent her to a fashionable girls' school. She hated the school and returned home after a year. But at least she could now read, and read she did, despite the complaints of relatives who saw that she was reading instead of sewing. It was by accident that she saw a problem in a magazine which was solved by algebra, and she wondered what this "algebra" was.

The tutor became sympathetic to her desire to learn mathematics . . . and helped her as much as he could. . . . Her father predicted that she would soon be in a straitjacket.

She wanted to buy a book on algebra, but it was not acceptable at that time for a young girl to go into a bookstore and buy such a book. She heard of Euclid's *Elements of Geometry* when her painting teacher recommended it to a male student, saying it would help him with perspective. Once more she could not obtain a copy because of her sex. Again, by accident, she was sewing in the room where a brother was being tutored in mathematics and, when he could not answer a question asked by the tutor, she prompted him. The tutor became sympathetic to her desire to learn mathematics, obtained a copy of Euclid for her, and helped her as much as he could. Her mother was horrified at her daughter's desire to study mathematics and, naturally, instructed the servants to take away her candles so that she could not study at night. Her father predicted that she would soon be in a straitjacket.

It was only after her second marriage to a cousin, William Somerville, a surgeon, that she found someone sympathetic to her desire for knowledge. There were now sufficient funds to allow her to follow her interests, but there were also four children. She managed somehow to find time to work. When the family moved to London, her husband's work brought them into contact with intellectuals. In this way she met the scientists of the day, among them the Herschels (Caroline, Sir William, Sir John), Sir Edward Parry, Lord Brougham, P. S. de LaPlace. Soon she was one of them. She presented her first paper, "The Magnetic Properties of the Violet Rays of the Solar Spectrum," to the Royal Society in 1826. In 1827 she was invited (by letter to her husband, asking him to persuade her) by the Society for the Diffusion of Useful Knowledge, to write a popularization of LaPlace's *Mécanique Céleste*. She accepted the invitation and translated the text, adding her own extensive notes. She called the work *The Mechanisms of the Heavens*. It was her most popular work, but she also wrote *The Connection of the Physical Sciences*, 1858; *Molecular and Microscopic Sciences*, 1869; *Physical Geography*, 1870, as well as many articles and monographs including one on "Curves and Surfaces of Higher Order."

Somerville took the view that mathematical truths existed in the mind of the Deity and that humans could only discover them, not create them. However, she was not so theological in her outlook that she escaped the criticism of the Church. Dean Cockburn of York Cathedral denounced her, by name from his pulpit, for her support of science.

All her life Somerville was an advocate of education for women and in her later years she wrote, "Age has not abated my zeal for the emancipation of my sex from the unreasonable prejudice too prevalent in Great Britain against a literary and scientific education for women."

In 1879, seven years after her death, Somerville College was founded at Oxford University as a women's college, which it still is today. In the last few years when three of the five women's colleges at Oxford have become coeducational, the Somerville College faculty has voted to keep the faculty, as well as the student body, all female.

In Russia in the 1800's encouragement for women to become mathematicians was not any greater than it had been in France in the 1700's. Sonja Corvin-Kurkovsky Kovalevsky (1850-1891) was born in Russia in 1850, the daughter of an artillery officer in the Russian army. In her teens she developed intellectual interests and was particularly fond of algebra. Her father's concept of a woman did not include her being an intellectual. When he learned that she liked algebra, he immediately threw away her algebra book. However, she was able to obtain another and continued her study secretly. Both she and her sister wanted university educations, but at that time in Russia women were not admitted to the universities. They determined to go abroad, but this was not possible as long as they needed the permission of their parents. With the aid of intellectual friends of their own age in St. Petersburg, they arranged a fictitious marriage for Sonja, when she was 18, with Vladimir Kovalevsky. The three set out for Heidelberg, the couple acting as chaperone for the other sister.

However, in Heidelberg the story was the same; women were not admitted to the University. After much effort on her part she was allowed to audit lectures. It was at this time that she decided that she definitely wanted to become a mathematician and that she wanted to work with Karl W.T. Weierstrass at the University of Berlin. So she went to Berlin, only to learn that opportunities there were even worse for women than in Heidelberg; women were not even allowed to audit classes. She wrote to Weierstrass and asked him to let her study with him as a private pupil. He sent her some problems to solve and was so impressed with her solutions that he consented to let her study with him. He soon came to consider her to be one of his most promising students. For the next four years she worked very hard.

In 1874 she finished three articles, any one of which, according to Weierstrass, was suitable for a Ph.D. dissertation. The question obviously was where she might receive a Ph.D. degree since it would be hard to receive a degree from the University of Berlin at which she had been denied admission. Weierstrass approached Göttingen, which had in the past awarded degrees in absentia. Finally, after many objections by Göttingen and efforts by Weierstrass, Göttingen awarded her the degree. In the one of her three articles which was used as a dissertation she solved a problem in partial differential equations posed by Cauchy. Her solution is now known as the Cauchy-Kovalevsky Theorem.

There followed a six-year period in which she did no mathematics; for reasons that are unknown, she returned to Russia where it was impossible for her to teach. She and her husband turned their fictitious marriage into a real one; she developed and followed other interests; she did not answer Weierstrass' letters. There were financial problems; her marriage deteriorated. After three years she decided to return to mathematics and asked Weierstrass for advice. However, she had a daughter soon afterwards and gave up mathematics to care for her. She soon tired of domesticity, and when P.S. Tchebyscheff invited her to give a paper at the Sixth Congress of Natural Scientists in 1880, she accepted, giving one of the three papers she had written in Berlin.

Gösta Mittag-Leffler heard her talk at this Congress, and offered to try to get her a position at the University of Helsinki where he was on the faculty. She decided against this, but did go abroad and began to work on mathematics again, sometimes in Paris, sometimes in Berlin. She left her daughter with a friend in Berlin. After her husband's suicide in 1883, Mittag-Leffler wrote Weierstrass that he had convinced the administration at Stockholm University, where he now was a professor, to allow Kovalevsky to lecture there, unpaid, of course. Finally, in 1884, the University offered her a five-year professorship.

She began working on a problem for which the French Academy of Sciences had offered a prize of 3000 francs, the Bordin Prize. The problem was that of determining the path of rotation of a solid body around a fixed point, where the path is contained in an ellipsoidal shell. The solutions were judged without knowledge of the identity of the authors. She received the prize, her solution having being so worthy that the prize money was increased to 5000 francs. Her memoir, "On the rotation of a solid body about a fixed point," for which she won the prize, is considered her most important work.

In the meantime she became emotionally involved with a Russian philosopher who was living in France. He proposed marriage on the condition that she give up her work. She refused. In 1889, at the end of her five year professorship at Stockholm University, she was made professor for life; in today's parlance, she received tenure. On return from one of her visits to her philosopher friend, she contracted influenza and died in Stockholm at the age of 41 and at the height of her mathematical powers. Shortly before, she had been elected to the St. Petersberg Academy of Sciences.

One might have hoped that opportunities for women mathematicians were better in the New World, but unfortunately they were not. Christine Ladd-Franklin (1848-1930), an 1869 graduate of Vassar College, wanted to be a physicist. She later learned that women were not allowed to work in laboratories and changed her interest to mathematics. When Johns Hopkins University, founded in 1876, announced a fellowship program in mathematics, her application, submitted under the name C. Ladd, was one of the first to arrive. Her credentials proved to be so outstanding that she

---

When the Board of Trustees discovered that she was a woman, they accused her of using trickery in order to gain admission and her fellowship was revoked.

---

was awarded a fellowship, sight unseen. When the Board of Trustees discovered that she was a woman, they accused her of using trickery in order to gain admission and her fellowship was revoked. Fortunately, this occurred while the world-famous English mathematician, James J. Sylvester, was at Hopkins. He had read some of her papers in English mathematical journals and insisted that the gifted young woman be admitted. The Trustees gave in, and Ladd entered Johns Hopkins in 1878 on a three-year fellowship. However, the Trustees forbade that her name appear in print in any list of fellowship holders at Johns Hopkins. In 1882 she submitted her dissertation, "The Algebra of Logic," which her adviser, Charles Sanders Peirce, said was brilliant. That was not good enough for the Trustees, however, and they ruled that no Ph.D. should be granted to her on the grounds that a precedent might be set.

Soon after this the Franklins left Johns Hopkins and went to Göttingen to study. There they found that women were not allowed to attend lectures. However, a member of the Göttingen faculty was so impressed with Ladd-Franklin's abilities that he gave his lectures to her privately and let her work in his laboratory. Out of this period came the beginning of her work on color vision, now known as the Ladd-Franklin theory. In 1904 she and her husband returned to Baltimore, he as editor of a Baltimore newspaper and she as a lecturer in logic and psychology at Johns Hopkins, the only woman member of the faculty. (Apparently, lecturers could be appointed without permission of the Trustees.) Still the Trustees refused to grant her the Ph.D. degree. In 1909 when Columbia University invited her to join its psychology faculty, she accepted.

It was while she was at Columbia that a member of the Psychology Department at Harvard University invited her to give a lecture there. She had accepted, and he had made all arrangements for her talk which was to be followed by a dinner in her honor, when Harvard's President heard of it. He immediately wrote to the psychologist who had invited her, saying that no woman was to speak at Harvard and that he must withdraw the invitation. She replied that, unless she heard directly from the President of Harvard, she was coming anyway. It seems that the President did not write, and Ladd-Franklin came to Harvard. Plans for her talk and visit went off as originally scheduled.

Finally, 44 years after Christine Ladd-Franklin had completed all the work for the Ph.D., and when she was 78 years old, Johns Hopkins awarded her the degree.

In the year in which Ladd-Franklin should have received her Ph.D., Emmy Noether (1882-1935) was born in Germany. Her father was the mathematician Max Noether, a professor at the University of Erlangen. Until she was 18 she seems to have followed the usual pattern for daughters of the bourgeoisie in Erlangen; she attended the State Girls' School, learned English and French, and took the Bavarian state examinations for certification as a teacher of those languages, should she ever need to earn her livelihood. It was soon after this that she decided she wanted to attend the University and to study mathematics. There are apparently no records left which might show why this young woman suddenly wanted to change the routine of her life and become a mathematician. At that time women were allowed to audit courses at the universities, providing they received permission from the professors. On this basis Noether audited courses at the University of Erlangen from 1900 to 1902. However, she wanted to enroll at the University as a regular student, which was, in general, not allowed. In fact, at that time women could not attend the Gymnasium which prepared male students for admission to the universities. But there was a loophole. Students could take a matriculation examination for admission to the university without having attended the Gymnasium. Noether took and passed this examination. Finally, she was permitted to enroll at Erlangen and in July, 1908 she was awarded her degree.

Noether's Ph.D. dissertation, "On complete systems of invariants for ternary biquadratic forms," was written under the direction of Paul Gordan. From 1908 until 1915 she stayed in Erlangen, without a position, doing research in mathematics, giving invited lectures, publishing papers and, toward the end of his life, substituting for her father at the University when he was ill. During this time she was beginning to be recognized as a mathematician of the first rank. In 1915 Felix Klein and David Hilbert invited her to come to Göttingen, which she did, staying until 1933 when she was forced to leave by the Nazis. Klein and Hilbert were particularly interested in her work on invariant theory for its usefulness in general relativity theory, on which they were working at the time. Yet however liberal Göttingen may have been in 1915 when compared to other German universities, it was still not ready for a woman faculty member. Despite the efforts of Hilbert and Klein, she was not granted a position which carried any stipend until 1923 when she was given a Lectureship in Algebra with a minimal stipend.

In the years between 1915 and 1923 she lectured at Göttingen, not under her own name, which was not allowed, since she was a woman, but under Hilbert's name. He announced lectures under his name and Noether delivered them. In her first years at Göttingen her research continued in invariant theory, resulting in two papers on differential invariants which are still used today. Noether's axiomatic approach to the study of abstract rings and ideal theory, for which she is best known, first appeared in a paper, written jointly with W. Schmeidler, in 1920. P.S. Alexandroff, the Russian mathematician, in his address to the Moscow Mathematical Society in 1935 at the time of her death said:

> Emmy Noether entered upon her wholly individual path of mathematical work in 1919-1920. . . . This work with W. Schmeidler serves as a prologue to her general theory of ideals, opening with the classical memoir of 1921, "Idealtheorie in Ringbereichen." I think that of all that Emmy Noether did, the bases of the general theory of ideals and all the work related to them have exerted, and will continue to exert, the greatest influence on mathematics as a whole . . . She taught us just to think in simple, and thus general, terms—homomorphic representation, the group or ring with operators, the ideal—and not in complicated algebraic calculations, and she therefore opened a path to the discovery of algebraic regularities where before these regularities had been obscured by complicated specific conditions.

In her work Noether stressed the use of chain conditions. It is because of her work that rings with the ascending chain condition on ideals are now known as Noetherian rings.

To quote Alexandroff again:

> From 1927 the influence of the ideas of Emmy Noether on contemporary mathematics continually grew, and along with it grew scientific praise for the author of those ideas. The direction of her work at this time moved more and more into the region of noncommutative algebra, the theory of representation and of the general arithmetic of hypercomplex areas . . .
>
> Emmy Noether at last received recognition for her ideas. If in the years 1923-1925 she had to demonstrate the importance of the theories that she had

developed, in 1932, at the International Mathematical Congress in Zürich, she was crowned with the laurel of her success. A summary of her work read by her at this gathering was the real triumph of the direction she represented . . .

It is interesting to note that both Noether and Germain did their most significant work relatively late for a mathematician.

According to Hermann Weyl, who became a member of the Göttingen faculty in 1930, Noether was "the strongest center of mathematics activity" at Göttingen from 1930 to 1933, "considering both the fertility of her scientific research program and her influence upon a large circle of pupils." Many of her younger colleagues, among them Weyl, recognized the injustice done to her in her lack of a suitable position, and tried to get a better position for her. In 1935 Weyl wrote:

> When I was called permanently to Göttingen in 1930, I earnestly tried to obtain from the Ministerium a better position for her, because I was ashamed to occupy such a preferred position beside her whom I knew to be my superior as a mathematician in many respects. I did not succeed. Tradition, prejudice, external considerations, weighted the balance against her scientific merits and scientific greatness, by that time denied by no one.

In 1933 with the Nazis in power in Germany, Noether, a Jew, was dismissed from her post at Göttingen. Alexandroff attempted to get her an appointment at the University of Moscow, but Russian red tape was slow and, when Bryn Mawr offered her a one-year, visiting appointment she accepted it. From Bryn Mawr it was an easy trip to the Institute for Advanced Study, and she soon began to give weekly lectures. At the age of 51 this remarkable mathematician held her first position at what might be called a "normal" salary. It is to the great credit of Anna Pell Wheeler, Chairman of the Mathematics Department at Bryn Mawr College at the time, and to the College itself, that Noether was invited to this country. In her second year at Bryn Mawr she entered the hospital for a routine operation to remove a tumor. She seemed to be recovering well, when suddenly she died, still at the height of her mathematical powers.

A legacy of barriers

The barriers which stood in the way of these six women, and the struggles they underwent, were tremendous. Even after their abilities were recognized by some enlightened mathematicians, they encountered severe difficulties in obtaining what male mathematicians, of equal or lesser abilities, obtained without question: appropriate degrees, appropriate positions with appropriate pay, and recognition by other male mathematicians as mathematical colleagues. These women from six different countries—Italy,

France, Russia, Great Britain, the United States, and Germany—faced almost identical problems over a time span of almost 200 years. In most cases each woman was aided by a broad-minded, liberal mathematician who recognized her ablity, strove to obtain for her some recognition in, and entrance to, a field marked "For men only." What each woman had within herself was the stamina and drive to continue in her chosen field, in spite of all the obstacles against that continuation. It is not surprising, then, that there have been few outstanding women mathematicians; what is surprising is that there have been any.

Even the history books have been unkind. E.T. Bell, in his much read *Men of Mathematics* (note the title) devotes one sentence and one footnote to Noether, who died two years before the publication of his book. Germain is mentioned (on four pages) in the chapter on Gauss; Kovalevsky is included in the chapter on "Master and Pupil." Agnesi, Somerville and Ladd-Franklin are not mentioned. In D.E. Smith's and J. Ginsburg's *A History of Mathematics in America before 1900,* on pp. 193-194, under "Interest Shown by Foreign Journals" they report that volumes 41-50 (1893-1897) of *Mathematische Annalen* contain 15 articles by Americans "including the following," of which 14, all by men, are listed. The only one omitted is by Mary Frances Winston Newson, a student of Felix Klein, who received her Ph.D. from Göttingen in 1897.

By the 1920's women in the United States were receiving about 6% of the doctorates awarded in mathematics, a figure that remained about the same until the beginning of the 1970's. Women could attend graduate schools at all of the universities with good research departments in mathematics except for one, Princeton University, which did not admit women to its graduate program in mathematics until the late 60's. However, in at least one of these institutions, Harvard University, although women graduate students were admitted, when they arrived they found that some professors did not allow them to sit in the classrooms, but did allow them to sit in chairs placed just outside the classroom doors so that they could hear the lectures. Until the 1970's, as far as employment was concerned, the situation remained much as Anna Pell Wheeler had described it in a letter written in 1910, soon after she received her Ph.D. degree from the University of Chicago:

> I had hoped for a position in one of the good universities like Wisconsin, Illinois, etc., but there is such an objection to women that they prefer a man even if he is inferior both in training and research. It seems that Professor E.H. Moore has also given up hope for he has inquired at some of the Eastern Girls' Colleges and Bryn Mawr is apparently the only one with a vacancy in Math.

(As an aside, it is worth noting that Anna Pell Wheeler was the first woman to be invited to give the prestigious American Mathematical Society Colloquium Lectures. That was in 1927. In 1980 the Colloquium Lectures were given by the second woman invited to give them. She is Julia Robinson.) Women mathematicians found employment mainly at the women's colleges and small liberal arts colleges where salaries were low, teaching loads were heavy, and where there was little or no money for help with paper grading. Seldom were there funds for sabbaticals or for travel to professional meetings. Women members of the faculty at these institutions were often required to perform duties having no connection with teaching, for example, living in the dormitories, acting as advisors to students and student groups, and even teaching Sunday School. The result was that there was little time for research and little money for travel to centers of mathematical research.

A decade of beginnings

At the beginning of the 1970's, whether spurred by the civil rights movement or the women's movement, some women mathematicians began to talk openly about discrimination faced by women in the mathematical community. For example, women mathematicians and women graduate students in mathematics in the Boston area were meeting to discuss common problems and possible solutions. In January 1971, at a meeting of the American Mathematical Society in Atlantic City, Mary W. Gray of American University and a group of women attending that meeting decided to form an association with the purpose of improving the status of women in the mathematical community to ensure that women will have the same opportunities in mathematics as men. This organization, the Association for Women in Mathematics (AWM) [2], elected Gray as its first president. The organization was at first small and operated as a grass roots movement holding local meetings all over the country. Now it has grown to approximately 1100 members, both male and female. The organization holds winter and summer meetings in conjunction with the joint meetings of the American Mathematical Society (AMS) and the Mathematical Association of America (MAA). At its January 1980 meeting the Association sponsored the first Emmy Noether Lecture, a lecture to be given annually by a woman mathematician. Regional meetings continue to be held throughout the year. AWM sponsors a Speakers' Bureau through which schools, colleges and universities can obtain women speakers on technical or non-technical subjects appropriate to a variety of audiences, from high school students to research mathematicians. In addition, a newsletter is published six times yearly with articles of interest to women as well as advertisements for positions in industry, government or academia. AWM was one of a collection of women's organizations which filed suit against the National Institutes of Health, challenging various agency procedures. Partly as a result of this suit, the processes have been opened up to wider participation and scrutiny, important features of the guarantees of due process.

The Association has encouraged mathematical organizations to consider both female and male mathematicians for candidates for offices and membership on committees, to open their nominations procedure to allow for nomination by petition, to invite more women to give hour talks and to organize sections on special topics at their meetings, and to increase the number (often from zero) of women editors on mathematical journals. Probably one of the most valuable contributions that the AWM has made to women mathematicians is to ensure that those who face discrimination in finding positions, in salaries, or in promotions, know that there are those who care, who understand their predicament, and who will try to help.

In 1971 the AMS Council authorized the President of the Society to appoint a Committee on Women in Mathematics "to identify and to recommend to the Council those actions which, in the opinion of the committee, the Society should take to alleviate some of the disadvantages that women mathematicians now experience."

The first task undertaken by the Committee on Women was a survey [10] of women who hold Ph.D. degrees in mathematics, which resulted in the publication of a *Directory of Women Mathematicians*. The *Directory* contains an abbreviated curriculum vitae for each woman (address, field, publications) which can be used by employers, committees seeking speakers, and officers seeking members for committees. The Committee also made a survey of the graduate school origins of female Ph.D.'s [6]. In addition to intrinsic interest in the information gathered, the Committee felt that it could serve as a guide to prospective female graduate students, and that it could provide each department with comparative information concerning enrollments of female graduate students. A study [5] of the effect of affirmative action on the composition of doctorate granting mathematics department faculties over the three academic years 1973-74 through 1975-76 showed that, contrary to the belief of many in the mathematics community, there had been no influx of women faculty members into these departments. Data used in this study came from the annual AMS surveys of mathematics departments. Due to a recommendation of the Committee the AMS surveys now report not only on the number of women in the mathematics departments at four-year colleges and universities but on the number on tenure. In the latest survey [1] the number of women on tenure in 1979-80 at the institutions in Group I, the 27 leading departments of mathematics in the U.S., was given as 21, the number of men on tenure as 655.

A subcommittee of the Committee on Women has met with representatives of the National Science Foundation to urge that the Foundation place greater emphasis on increasing the number of women in mathematics and science. The subcommittee urged the Foundation first to review its own hiring practices to ensure that women scientists are employed at the NSF. The subcommittee suggested that the NSF establish special programs for young researchers, any of which would benefit women. These could include an increase in the number of postdoctoral fellowships; grants to academic institutions to be used by the institutions to reduce teaching loads, to allow more time for research; or travel grants for attendance at professional meetings. The Committee on Women is currently seeking support for a program of visiting women lecturers and consultants in mathematics. As many academic institutions in the country, among them most of the large universities with the most prestigious graduate schools, still have no women on the mathematics faculties, or certainly none on the tenure track, the Committee hopes that these visits would give a wider audience to the work of women mathematicians. The Committee also hopes that women students at these institutions, by seeing women mathematicians acting on an equal basis with their male colleagues, would be encouraged to continue in mathematics.

Many academic institutions in the country, among them most of the large universities with the most prestigious graduate schools in mathematics, still have no women on the mathematics faculties, or certainly none on the tenure track.

As time went on other mathematical organizations joined in sponsoring the Committee on Women in Mathematics so that by January 1980 the Committee was a Joint Committee of the AMS, the MAA, the National Council of Teachers of Mathematics (NCTM), and the Society for Industrial and Applied Mathematics (SIAM).

Due to their historical backgrounds, both the AWM and the AMS-MAA-NCTM-SIAM Committee on Women in Mathematics have focused mainly on the problems of college women students and women who hold advanced degrees in mathematics and who, for the most part, are employed in colleges, universities, government, or industry. But discrimination against women in mathematics has not been limited to women already in college or to those holding advanced degrees. From elementary school through high school, girls have often not been encouraged to continue the study of mathematics; in many cases they have been actively discouraged [4]. For a discussion of these issues see the article "The Real Energy Crisis" by Eileen L. Poiani elsewhere in this volume.

By the middle 70's a few programs designed to combat this problem were being developed. On a national level, the MAA started a Women and Mathematics Program (WAM) of visiting women lecturers to secondary schools. In 1978, at a national meeting of the NCTM, a group of women formed an organization, now known as Women and Mathematics Education (WME) [11], dedicated to promoting mathematics education of girls and women. During 1977-79 the National Science Foundation sponsored a Visiting Women Scientists Program for junior and senior high schools, where each woman scientist spent a day at a school. The Foundation hoped that these women would act as role models for the girls and encourage them to study science and to seek careers in the sciences.

On a local or regional basis quite a few programs have been established with the aim of educating girls, women, teachers and counselors to the importance of mathematics in nearly all fields today and of encouraging girls and women to enter careers in mathematics or the sciences. One of the most successful of these has been the Math/Science Network [8] of the San Francisco Bay Area, which started in 1973 with women mathematicians, scientists, engineers and educators developing such programs. The Network, working under a grant from the Carnegie Corporation, covers both pre-collegiate and collegiate programs, with Nancy Kreinberg of the Lawrence Hall of Science at the University of California at Berkeley as Director of the pre-collegiate program, and with Lenore Blum of Mills College (a former President of the Association for Women in Mathematics) as Director of the collegiate program. Among the many programs of the Network are classes for girls 6 to 12; conferences for junior and senior high school girls; special programs for women in science, including one for mature women; the Equals program in teacher training; and the development of several films on women in science. The films and the "Equals" materials are being distributed across the country.

In the 70's many colleges and universities developed special mathematics courses and programs for women students. Wellesley College, a women's college, and Wesleyan University, a coeducational institution, were two of these institutions. Both received outside funding for their programs: Wellesley in 1975-76 from the Alfred P. Sloan Foundation, Wesleyan in the same year from the Fund for the Improvement of Postsecondary Education (FIPSE); for the next two years the two shared a grant from FIPSE. Wellesley's program contained several components: the development of a one-semester course, A Discovery Course in Mathematics and its Applications; two short noncredit courses, a precalculus one and one devoted to preparing juniors and seniors for the many professional examinations; and a psychological evaluation of the impact of the "Discovery" course on the students who took it. These were students who had dropped the study of mathematics before entering college and who had planned to take no mathematics courses or courses requiring quantitative techniques in college and who had been persuaded to take the course.

The "Discovery" course is taught as a discussion course in which the students "discover" mathematical theories and techniques: the instructor asks one or two questions from time to time and then leaves the students to

develop the material themselves. There is no text. The psychological evaluation was never a part of the classroom activity. The psychologist held
individual meetings with the students or used questionnaires given to the
students at the beginning and end of the course. One of the most interesting
results occurred in a semester when 53 applied for space for 15, so that a
good control group was available. In answer to a question about the grade
expected in a physical science course, should the student take one, at the
beginning of the semester the majority of both the control group and the
group taking the course answered D; at the end of the semester the control
group still answered D, but the class group now answered B.

At Wesleyan the program was very different. There a Math Clinic
operated, with psychological support to try to alleviate the fears that many
students, both male and female, feel about mathematics and to help advise
them of appropriate mathematics courses. In addition, a course in intuitional calculus, interlaced with a review of algebra as needed, was offered
to a large group of students as a lecture course with recitation sections.

Programs similar to those mentioned above, or adaptations of them,
have been developed in many areas of the country. It is impossible here to
mention all of them. The ones mentioned above are, in many ways, typical
of them all.

Current status

With all the activity of the past decade, one wants to say that women are
now treated on the same basis as men in the American mathematical
community. Unfortunately, this is not yet the case. In a report on the status
of women in the sciences prepared for the National Reseach Council [7]
and published in 1979, it was recorded that at the 25 institutions in the
country which receive the largest amounts in federal expenditures for
research and development the number of women on the mathematics
faculties at the rank of associate or full professor was 13 in 1975-76, while
the number of men at those institutions in those ranks was larger than 500.
Since 1976 the number of women at those ranks at those institutions has
increased by less than 5.

The AMS annual survey on the status of the profession (faculty salaries,
tenure, women, new doctorates) published in October, 1979 [1] shows that
13.7% of the new U.S. 1978-79 doctorates were received by women. This is
slightly more than the 13.3% reported for 1977-78, and continues the
increase since 1970 when, as previously cited, approximately 6% of the new
doctorates were granted to women. The same survey shows that of those
holding new doctorates employed in the mathematics departments in the
U.S., 10% are women, but no breakdown is given for the percentage
employed at the 65 leading departments. However, the survey reports that

among new doctorates employed by bachelors and masters degree-granting departments, 23% are women, while among those employed by government, business and industry, only 14% are women. The next survey should include a report on the 65 departments. Among those holding new doctorates and unemployed at the time of the survey, 21% were women.

The same survey shows that on the mathematics department faculties of the doctorate granting institutions in the U.S. in 1979-80 the number of doctorate-holding women on tenure is 2.5%, while the number of doctorate-holding men on tenure is 70.6%. Of the women employed on these faculties, 47.9% are on tenure compared to 74.5% of the men. Comparable figures for women and men not holding the Ph.D. degree are 15.4% of the women on tenure as compared to 42.7% of the men; the percentage of women employed on tenure is 52.1% as compared to 60.6% of the men. These figures indicate that women are not being employed at the prestigious institutions in proportion to the number who hold Ph.D.'s, that the number on tenure is lower than that for men, and that women seem to be obtaining tenure slower than men. It is fair to conclude that women graduate students in mathematics are still seeing very few women faculty members and even fewer on tenure. The situation is somewhat brighter at the bachelors and masters granting institutions. However, to end this paragraph on a happier note, at the five departments considered by mathematicians to be the leading ones in the country, there were no tenured women on the faculties in 1970; now there are three.

Betty M. Vetter, Executive Director of the Scientific Manpower Commission, pointed out, in a document dated March 3, 1980, that of the individuals surveyed in 1977 who held Ph.D.'s in mathematics, the unemployment rate for men was 0.9% and for women 3.6%. The average salaries for these people in that same year differed by approximately $4000 ($23,000 to $19,000). For those receiving bachelor's degrees in mathematics in 1976 and surveyed in 1978, the unemployment rate for men was higher than that for women: 3.6% to 2.0%; on the other hand, for those graduating in 1978 the job offers to men averaged $1,192/month compared to $1,177 for women. By 1979 the latter figures had grown to $1,340 for men to $1,304 for women. She goes on to report:

> It is not only in the academic world that women scientists and engineers are treated differently than men. They are seriously underrepresented in industry in comparison with their availability. ... In the government, women scientists and engineers are employed in somewhat closer approximation to their proportions within the available population, but they are generally employed an average one and a half civil service grades below men with the same credentials. ... The fact that women advance more slowly in the academic ranks as well as in government and industry accounts for much of the salary gap that persists into the present, and has in some instances, widened. But the gap begins with the first job after college. Except for new baccalaureate

engineers, women earn less than men in every science field, at every degree level, at every level of experience and in every employment sector. The salary gap widens with age and experience, and with higher degree levels.

Despite the inequalities still existing for women in the employment situation, there have been some improvements in the last decade. More women among the new doctorates are being employed at the university level than previously, some of them on special appointments of two or three years which carry reduced teaching loads and so more time for research. Some few women are receiving tenure at universities, a few more at coeducational four-year colleges.

At the end of the 70's one can see some progress in other areas, too. The Sloan Fellowship Program for research mathematicians less than 32 years of age has included its first woman, and several more since the first one. The National Academy of Sciences has elected its first woman mathematician, unemployed at the time of her election and almost immediately offered a professorship at the University of California at Berkeley following the election. A woman, Mary Ellen Rudin, has delivered the Earl Raymond Hedrick Lectures on invitation by the MAA; there has been an increase in the number of women invited to give one-hour lectures at both the AMS and MAA meetings. There have been two women vice presidents of the AMS; the MAA now has a woman president, the first in its history. The number of women editors on 25 mathematical journals has risen from 9 of 459 editors in 1977 to 19 in 1980. High school teachers report an improvement in some of the newer textbooks in their depiction of boys and girls, getting away from the stereotype of one set of careers for girls and another for boys. At some high schools just about as many girls as boys are taking the Advanced Placement examination in mathematics for entrance to college [3]. So there is some progress, albeit slow.

Goals for the 1980's

Clearly, the primary goal for the 80's should be an improvement in the employment situation, where men and women, with equal qualifications, have equal opportunities in the job market; it should not be necessary for a woman to be a superstar to compete with a man who is well below the rank of superstar. If colleges and universities, businesses, industries and government agencies are not employing women, then there should be viable programs to ensure that they are. Academic departments should not be satisfied, nor allowed to be satisfied, with the token woman member on tenure. Nor should any business, industry, or government agency which employs mathematicians, be satisfied with employment of the token woman. Women with the same qualifications as men should be allowed to go up the "ladder" at the same rate as men and with equal salaries.

Early in the decade a program of women lecturers and consultants in mathematics, similar to the one proposed by the AMS-MAA-NCTM-SIAM Committee on Women in Mathematics described earlier, should be funded. Such a program would serve to demonstrate to mathematics departments and, especially to those with few or no women members, and to the students in those departments, that women are as capable as their male colleagues.

There should be an increase in postdoctoral fellowship programs for recent Ph.D.'s. Any fellowship program for people who have held doctorates less than, say, five years would benefit women. However, because most women in this age bracket are of child bearing age, the five year period should be extended to seven years, a policy now in effect in some programs for applicants who have served in the military forces. There is, moreover, a real need for a flexible research support program for recent Ph.D.'s in which grants would be given not only to faculty members but to individuals without institutional affiliations, and in which there would be part-time research grants. Travel grants for recent doctorates which would cover expenses for attendance at professional meetings or research conferences for those employed at institutions with little funding for travel would specially benefit women, because many women find employment at such institutions. A logical agency to lead the way in inaugurating these programs is the National Science Foundation.

But first the NSF should look at its own internal structure. It should check its recruitment procedures and hiring practices. In the latest listing [9] of the directors and heads of the various programs in the Division of Mathematical and Computer Sciences one woman's name appears as contrasted to those of 17 men. Women, not just one woman, should be appointed to all NSF advisory committees and task forces. The NSF should check its peer review system for proposals submitted to it to ensure that women are used as reviewers and that proposals submitted by women are reviewed fairly. These recommendations to the NSF are not new; for example, they were recommended in "The Participation of Women in Scientific Research," a report issued in March 1978 by the Office of Opportunities in Science and the AAAS. Indeed, all federal agencies concerned with the sciences at any level need to examine their policies to determine if they are discriminatory as far as women are concerned.

Another goal for the 1980's is early passage of the Women in Science and Technology Equal Opportunity Act, just passed in the Senate as Bill 568. When S 568 was introduced in the Senate on March 7, 1979, Senator Edward M. Kennedy of Massachusetts said:

> The National Science Foundation has described the underrepresentation of women in science and engineering careers as a "significant national problem." Nevertheless those in the highest policy-making positions have failed to assign any priority to the funding or support of efforts to examine, understand and change the present situation.

Some portions of the Women in Science and Technology Equal Opportunity Act were incorporated in the 1981 National Science Foundation Authorization and Equal Opportunities in Science and Technology Act signed by President Carter on December 12, 1980.

The "Women in Science" Act is designed to rectify some of the discriminatory practices against women in science. The bill calls for a five-year, $125 million program to "assure equal opportunity for women in education, training, and employment in scientific and technical fields." The program is to be carried out under the direction of the NSF, and calls for the cooperation of the industrial and academic sectors in accomplishing the goals of the bill. Programs to be financed, should the bill become law, cover all areas in education from elementary school through graduate school. Among the programs to be financed are fellowship and grant programs, a program to inform the public of the importance of the participation of women in science, and a visiting women scientists program. Very importantly, the bill also calls for the establishment of a Committee on Women in Science, composed of 13 members, at least nine of whom are to be women, of whom seven are to be doctoral scientists or engineers.

Another goal of the 1980's should be the further democratization of mathematical organizations. A good start was made in the 70's, but there are still changes that need to be made, not least, provision for nomination to all offices by petition. Blind refereeing should be tried again, possibly in two journals this time. In the 70's blind refereeing was tried in the *AMS Proceedings* for two years and then abandoned. Perhaps it is worth recalling that Sonja Kovalevsky won the Prix Bordin of the French Academy of Sciences with a paper submitted anonymously. Of course, there are obvious differences between reading a set of papers in a prize competition and reading papers submitted to a journal. On the other hand, if some individuals have felt discriminated against in the past because their papers carried a clearly feminine name, how can anyone object to blind refereeing? Due to the work of the AWM there has been an increase in the number of women editors of mathematical journals, but the increase is minimal. Perhaps the American Mathematical Society might celebrate its 100th birthday in 1988 with its first woman president!

What should be the goals for the 90's? Let us hope none will be needed.

References

[1] Twenty-Third Annual AMS Survey. *Notices of the American Mathematical Society* 26 (1979) 382-392.
[2] Association for Women in Mathematics. Women's Research Center, Wellesley College, Wellesley, MA 02181.

[3]   Patricia Casserly. "Helping Able Young Women Take Math and Science
      Seriously in School." *New Voices in Counseling the Gifted*. Ed. by Nicholas
      Colangelo and Ronald Zaffrann. Kendall/Hunt, 1979.
[4]   Mary W. Gray. "The Mathematical Education of Women." *American Mathe-
      matical Monthly* 84 (1977) 374-377.
[5]   Mary W. Gray and Alice T. Schafer. "Has Affirmative Action Affected the
      Composition of Doctorate Granting Mathematics Department Faculties in
      the USA?" *Notices of the American Mathematical Society* 23 (1976) 353-356.
[6]   I.N. Herstein. "Graduate Schools of Origin of Female Ph.D.'s." *Notices of the
      American Mathematical Society* 23 (1976) 166-171.
[7]   Lilli S. Hornig. *Climbing the Academic Ladder: Doctoral Women Scientists in
      Academe*. National Academy of Sciences, Washington, DC 20418, 1979.
[8]   Math/Science Network. c/o Lenore Blum, Mills College, Oakland, CA
      94613.
[9]   *Mathematical Sciences Administrative Directory*. American Mathematical Soci-
      ety, Providence, RI 02940, 1979.
[10] Cathleen Morawetz. "Women in Mathematics." *Notices of the American
      Mathematical Society* 20 (1973) 131-132.
[11] Women & Mathematics Education. c/o Joanne Becker, School of Education,
      2303 James Hall, Brooklyn College, CUNY, Brooklyn, NY 11210.

Spatial Separation in Family Life: A Mathematician's Choice

Marian Boykan Pour-El

The media have recently been reporting some strange stories. Are we witnessing the next step in the Women's Movement? Married couples are living apart in order to practice their careers. The papers seem to assure us that these couples intend to do this for at most a year or two—and not at all if children are involved. Interviews with husband and wife reveal that both believe they must live apart if they are to succeed professionally; the job market is tight. I sometimes wonder whether these couples realize what might be in store for them.

For approximately twenty-three years, my husband and I have alternated periods of "living apart" with "living together" in order to pursue our careers. During that time we raised a daughter from a two-year-old toddler to a twenty-five-year-old college graduate. For this reason I have often been called upon to lecture and advise about "life apart" as a basis for a two-career family. Although I would prefer anonymity in this matter, I usually consent to give whatever help seems necessary. Contrary to the prevailing attitude, our family believes that our life was richer for having lived apart than it would have been had we lived together continuously. Furthermore our experiences indicate that continuous physical nearness need not always be a primary consideration in family life. If a mother's love for her grown children can extend undiminished over the miles, why cannot other aspects of family life survive and flourish under similar conditions?

Marian Boykan Pour-El is Professor of Mathematics at the University of Minnesota. She received her Ph.D. from Harvard University in 1958. From 1962 to 1964 she was a member of the Institute for Advanced Study in Princeton, New Jersey. Her research interests are primarily in mathematical logic and theoretical computer science. She has delivered invited addresses to the American Mathematical Society and also to international congresses specializing in these fields. In addition, she has lectured on "living apart and the two-career family." At present she is a Trustee and Member of the Council of the Conference Board of the Mathematical Sciences, and a Member of the Council of the American Mathematical Society.

In this account, the focus will be on four distinct aspects of this type of marriage-arrangement: *long-term separation, short-term separation, commuting*, and *living together continuously*. These are explained later, accompanied by glimpses of the lifestyle. Further details, including comments on the texture of this way of life, appear as reflections at the end. But let us first turn to some background material.

Background

Our lifestyle would never have begun without my deep commitment to professional life. Hence its background is really an account of career development. For me it is an account of growing up in the rough and tumble of New York City, with very little support from anyone.

The decision to seek a career was made when I was very young. I did not know exactly what career I wanted. But with the melodramatic flair characteristic of a young child, I wanted to "understand the universe, live completely, and contribute to society." I was aware that my father, a dentist, believed that a woman could not possibly be good in his profession. (My gentle father would have kept his thoughts to himself had he known of my plans.) I knew that my mother, while chatting with neighbors, had expressed a desire to see her son graduate from college. Her daughter would go only if there was excess money. Nonetheless, with the characteristic optimism of the very young, I expected to find a way to a career. I expected to succeed.

My early schooling left much to be desired. If grade school was mediocre and uninspired, junior high was considerably worse. Armed gangs roamed the halls threatening those who interfered with their pleasure. To avoid attack, girls were not permitted to leave the classroom alone. Our teachers spent so much time and energy protecting themselves from gang wrath that instruction and guidance were at a minimum. In order to climb out of this morass, I was forced to take matters into my own hands. I did this by choosing a high school carefully.

My choice was Hunter High, a free college-preparatory school for women. I would have preferred Bronx Science but girls were not admitted at that time. Upon graduation I entered Hunter College, a free college for women, as a physics and mathematics major. I did quite well in school both scholastically and socially. I gave to my classmates what they gave to me—companionship. It did not matter that my friends were oriented towards marriage and I towards a career. I considered it a measure of success in my college friendships that I was chosen to be a bridesmaid at their weddings in spite of our differing point of view. This confirmed my belief that one did not have to espouse the same opinions, to bend to the dictates of society, to get along and have friends. That valuable lesson in mental nonconformity has served me well throughout my life.

In 1949 I passed from an essentially all female environment to one which was almost totally male. This happened quite suddenly, when I accepted a fellowship from Harvard Graduate School. Many amusing episodes about a woman placed in a male environment could be told. Here is one example. A few years after I arrived at Harvard a notice appeared on the bulletin board advertising an instructorship in mathematics at Yale. Inked in by hand by a member of the Harvard faculty were the words "For men only." As I was the only woman seriously attempting to obtain a Ph.D. in mathematics, my fellow students knew who was excluded. I would have had an excellent chance of obtaining the position had I been a man. As one of my friends happily expressed himself upon seeing the notice, "I am one up on you in this respect." What was my feeling about the matter? It was one of gratitude that the faculty had thought enough of me to ink in the addition. I had expended considerable effort at Harvard to overcome an inadequate mathematical background. This remark gave me a sense of achievement.

Little discriminatory acts such as professorial stag parties for graduate students—momentarily they forgot I existed—played a rather minor role in my life at Harvard. Of greater importance was the fact that, several years later, the Harvard faculty helped beyond the call of duty. I had disappeared from Harvard and, on my own, had written a Ph.D. thesis in mathematical logic. This was a specialty about which the Harvard faculty had little expertise. I had no thesis advisor, nor any thesis committee. Yet the Harvard faculty took it upon itself to have the thesis evaluated by authorities from outside. I was granted my Ph.D. in 1958, many years before the advent of the current Women's Liberation Movement. By that time I was married to a biochemist and had a young child; I was well-practiced in pursuing a non-traditional course.

A glimpse of the lifestyle

Our lifestyle evolved slowly, beginning as a temporary response to our professional needs. Although we had already committed ourselves to a two-career family before marriage, we never considered the possibility that we would not live together continuously. Our lifestyle, once begun, progressed on its own as we gained experience and understanding. We distinguish four separate components:

Long-term separation: apart for several months at a time, together for several weeks at a time.

Short-term separation: apart for brief periods of approximately one to four weeks, together for even briefer periods, usually a weekend.

Commuting.

Living together continuously.

Each of these was *practiced in connection with a two-career family*. As a consequence, our "short-term separation" differs from that of the travelling salesman, our "long-term separation" differs from that of the soldier or sailor. Furthermore each component was affected by other considerations —the age and needs of our child, our career demands, the place where we lived. In what follows I give a brief historical account including all four components and their interactions with the conditions affecting them.

It all began almost by chance in 1958, shortly after I obtained my Ph.D. I received an offer of a "tenure track" assistant professorship at Pennsylvania State University. As Penn State had some very fine mathematical logicians and no nepotism rule, I could not refuse. My husband, who had not quite completed his Ph.D., would remain in Berkeley, California. My daughter and I would move to Pennsylvania. Except for one brief visit of three weeks, we would live apart for an academic year. We would experience a *long-term separation*.

We planned our move very carefully, and with particular concern for our 2 1/2 year-old daughter. She was to attend nursery school in Pennsylvania as she had enjoyed one in California. Communication was to be by letter. We had no extra funds for long distance telephone calls or frequent visits. It all seemed safe and assured, if spartan. Unfortunately, Murphy's Law took over; nature intervened and my daughter contracted bronchitis. Central Pennsylvania was no sunny California. Although my daughter managed to catnap between bouts of coughing day and night, I could not. Imagine the situation. I was all alone—teaching, doing research, homemaking, and taking care of a sick child—with little sleep. The first attempt at our lifestyle (although we did not know then that it was the beginning of a lifestyle) seemed doomed to failure. However we had journeyed too far to turn back. Nursery school was out. After several months I found an adequate babysitter, no easy task in central Pennsylvania. Life could once again proceed on a more even keel.

My husband's single visit during that academic year was awaited with great anticipation. My daughter and I talked about it incessantly. When he arrived, the interlude of togetherness was more than we had hoped for. The bond between us had strengthened, our discussions were deeper and more meaningful. We parted with sorrow, but also with joy at the discovery of our strengthened ties. The long period of separation had provided the impetus for the heightened joy of togetherness. We did not realize then that we had had a glimpse of a new lifestyle: the yearning when apart, the delight when together. We were living in limbo, waiting to be reunited again.

Akiva, my husband, returned as planned in the summer with his Ph.D. and a position at Penn State. We remained together in Pennsylvania for three years. These were good years. By then I was promoted to associate professor. Although we did not realize it, this was but a prelude to other separations.

The next in the series was a *short-term separation*, begun when I received a fellowship to the Institute for Advanced Study in Princeton. Since the Institute was associated in my mind with Einstein, Gödel, and other mathematical luminaries, I could not refuse the offer. Akiva would remain in Pennsylvania and my daughter and I would live in New Jersey.

Again we planned carefully. My husband would visit us by car on alternate weekends—a seven hour trip, one way. Air and rail transportation to central Pennsylvania were virtually non-existent. As Ina was six, she needed a babysitter only for after-school hours. In Princeton, this was easy to find.

The weekends during which my husband visited were very satisfying. All of our activities were family affairs—none of the "you do this while I do that" arrangement. The intervals of separation provided enough stimulus so that strong bonds were forged during the periods of togetherness. Yet these visits did not have the power of those experienced during a long-term separation.

Our *commuting* experience was initiated by my husband a year later. During my second year at the Institute, Akiva developed a severe back ailment. After being stretched, pounded, injected and immobilized to no avail by several doctors in central Pennsylvania, he came east to join me and to secure adequate medical service. When he accepted a position at the Veteran's Hospital in Philadelphia, he began commuting. Although we found this to be a pleasant experience, it did not have the positive effect of even the short-term separation.

While in Princeton, we began to search for a place where we could live together continuously. We still had not realized that the patterns of alternating separation and togetherness were to continue throughout our lives. But we had already become quite experienced, both mentally and physically, in undertaking such an arrangement.

Our search led us to the Twin Cities where we *lived together continuously* for five years. During that time I became a Professor at the University of Minnesota and Akiva was promoted to Section Manager at Archer Daniels Midlands Research Division. Many times my lecturing took me away from home, not only throughout the United States, but also to European countries on both sides of the Iron Curtain. My daughter never accompanied me on these trips. However, the pattern of togetherness remained intact, partly due to the fact that each separation was quite brief and did not involve setting up a new household. To preserve our togetherness I refused several offers as visiting professor at other universities. We were living together continuously as we had anticipated at the beginning of our marriage.

The sudden realization of our lifestyle came in 1969. In that year my husband's laboratory moved lock, stock and barrel to Decatur, Illinois. Decatur may have its favorable aspects, but these seemed well hidden. All that was apparent were soy-beans, corn, railroad tracks, and a rather mediocre school system. Once again, a separation was necessary. It would be *long-term*, since transportation between Minneapolis and Decatur was very poor. This time we really understood the significance of what we were doing. We were opting for the ever changing variety of a "long-term separation" over the more even existence of "living together continuously." But we went one step further. If we were to have a long-term separation, could we not be a continent apart? I decided to go to Bristol, England, a center for mathematical logic. My daughter, when asked, elected to join me. If you had to choose between England and Decatur which would you pick?

The arrangments for our stay in Bristol were relatively easy to make. I planned to live quietly in Bristol. Ina was enrolled in school and I settled down to work. Unfortunately, informed of my presence through the grapevine, my colleagues in England and the continent invited me for lectures and short term visits. This surprising development demanded some new arrangements for my thirteen-year-old daughter. Had I been a man, it would have been so simple: leave the child with her mother. Babysitting as we understand it in the U.S.A. was unknown. My colleagues in Bristol were at loss to advise me: this was new to them too. By chance I discovered an English family willing to take Ina in for weeks at a time. We tried this arrangement on some of my overnight trips within England, e.g., to Oxford. When it became clear that Ina was enjoying life in an English household, I was able to plan trips to the continent, each several weeks in length. Thus I was able to keep my commitments both to colleagues and family. The latter would have been easier if I had had more than one child.

Ina's experience with an English family had a profound effect on her. She had arrived in England, a timid thirteen-year old, afraid to use English money. She left, as she herself had wished, by flying alone to Paris to a French summer camp. She had become a confident fourteen-year-old.

During that year abroad, my husband visited us twice. He came during Christmas, about five months after we had parted; he returned during the summer. Each visit was three weeks in length. In each we experienced the joy we had come to expect of the long-term separation.

After fourteen months abroad, my daughter and I returned to Minnesota; my husband remained in Illinois. We continued this long-term separation for the next five years. Very recently my husband has left Illinois to do consulting work in biochemistry. This takes him to many distant places for varying lengths of time. Today long-term separations as well as short-term separations alternate with periods of living together.

When our daughter left home to go to college, our lifestyle expanded to include her as an *equal partner*. "Up to now," she said, "I have been following one of you. Now we each have our own center." To her it seemed a natural thing to do once one reaches adulthood: a physical separation which does not connote an emotional one. At times all three of us are apart, at times two of us are together, and sometimes all three of us are in the same place. We are still practicing the lifestyle we began twenty-three years ago.

Reflections

As my husband and I review our married life over the past twenty years we find that we are in agreement about one fundamental point: life was richer than it would have been had we lived together continuously. Let me attempt to explain this.

There are two major reasons for our point of view. First, our lifestyle paved the way for increased mutual respect. Living apart made it possible for us to practice our careers and achieve professional recognition. Even when apart, we felt that each had a worthwhile occupation and each was leading an interesting life. Thus each became a more fascinating person to the other. Second, our lifestyle helped us acquire increased inner strength and independence—two qualities which we both prized. I did not need my husband to support me. I was doing that on my own. My husband did not need me to sew, cook or clean for him. He had sharpened these skills during the long-term separations. We did not remain married because we feared loneliness. We knew how to live alone successfully. We did not fear bringing up a child alone. We were doing that too. We stayed together because we wanted to and not because we felt we must in order to survive. Our separations forced us to strip our lives of those aspects of family living which were unessential to us and focus on those we considered important.

The third reason is concerned with interchangeability and equality in family roles. This happened, not only when we were apart, but also when we were together. As we had so little time together, only matters of importance to us were included in our plans. Unimportant matters, such as household chores, rarely concerned us. We attempted to discharge as many of these as possible while alone. Interestingly enough those which remained always got done. In the spirit of good will and helpfulness which always accompanied our meetings, one of us—although not always the same one—would finish off the job as quickly as possible so that we could go on to other things. Usually this happened without determining in advance who would do it. This made our roles within the family more fluid. There was little time to stake out a position, either a traditional one or one motivated by the Women's Movement. There was even less time to enforce that position. Nevertheless we did achieve some of the same results that other couples, influenced by the Movement, obtain by negotiation, argument, compromise and marriage contracts. Ours came easily and naturally as a consequence of our lifestyle.

The fourth reason is one of sheer enjoyment. Our lifestyle provided an element of newness, of dynamism, of change, and of excitement. The long-term separations gave us enough time and distance to compare life apart with life together. We were continually evaluating and re-evaluating our marriage. When apart we enjoyed lives which were emotionally and intellectually stimulating. We thought of each other constantly. When we met after a long-term separation it was not the "cocktail hour" or the ritual "evening meal" type of togetherness we remembered from the days when we lived together continuously. Discussions were considerably deeper, emotional investments considerably stronger.

In recent years our lifestyle has helped us in a fifth way. My women friends with grown children have experienced the "empty nest syndrome." When their children left home they felt somewhat lost, unneeded and unwanted. This was true even of those who were professionally active as doctors, lawyers, etc. I never shared this feeling. This is not surprising. Our marriage was not based on continuous physical nearness. Hence our daughter did not leave the family circle simply by living far away from her parents.

Let me make it very clear. There are some difficult problems connected with this lifestyle. One has to work hard to achieve some of the effects mentioned above. A full discussion of this would require a separate article; instead I will consider briefly two sample problems. The first is the matter of sex. Clearly a lifestyle such as ours does not permit sex every other night —or even sex every week. Some agreement between husband and wife appears necessary. Our solution was abstinence while apart. Although it may not be essential to the success of the lifestyle considered abstractly, we consider abstinence important to us. Is this our link to tradition?

A much more subtle problem is the matter of socializing in a "couple-oriented" society. I believe that this is one of the most important and delicate problems confronting the participants of a non-traditional lifestyle. Unfortunately, lack of space prevents me from describing its dynamics. For the present let me merely state that, even when alone, my husband and I are each invited to parties, theatre outings, and other social events on a regular basis. It is easy and natural for us to participate, and we usually do. In brief, my husband and I lead active social lives when apart as well as when together.

More than forty years ago, I decided to have a career. I was amused when told that a full-time mathematical career was incompatible with successful family life. The Women's Movement of the late sixties changed all this. The career-wife is now in fashion.

More than twenty years ago we began this lifestyle. Today I am again amused when told that what we did is incompatible with successful family life. Ours is richer for having lived apart. It is my hope that those who are beginning to practice this lifestyle, perhaps with considerable trepidation, will be somewhat reassured by our experiences.

Mathematics for Tomorrow

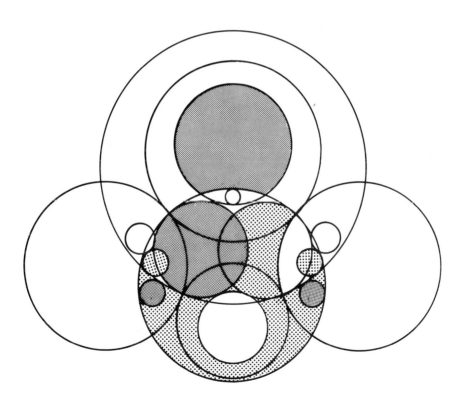

Applications of Undergraduate Mathematics

Ross L. Finney

In recent years there has been a phenomenal growth in the professional use of mathematics, a growth so rapid that it has outstripped the capacity of many courses in our schools and colleges to train people for the mathematical tasks that are expected of them when they take employment. People who take jobs with the civilian government, the military, or industry, or who enter quantitative fields as graduate students or faculty, discover with increasing frequency these days that they lack acquaintance with important mathematical models and experience in modeling. Many of them also find to their distress that they have not been trained to be self-educating in the application of mathematics.

This discovery, perhaps I should say predicament, is not the exclusive domain of people who enter fields that depend for their progress upon advanced mathematics. In Louisville, Kentucky, the profession of interior decorating is highly competitive. To stay in business, a decorator must be able to make accurate cost estimates. To do so without delay requires facility with decimal arithmetic, fractions, and area formulas. People hired as stenographers by The First National Bank of Boston discover that the work is done not on typewriters but on computer-driven word processors. Many stores now use their cash registers for inventory control. The keys on business machines have multiple functions, and the functions must be

Ross L. Finney is currently Senior Lecturer at MIT and Project Director of the Undergraduate Mathematics Applications Project at Educational Development Center, Inc. He was a Fulbright Scholar at the Poincaré Institute in Paris, France, in 1955, and earned a Ph.D. in mathematics from the University of Michigan in 1962. He has taught at Princeton University and at the University of Illinois at Urbana-Champaign. From 1962 to 1970 Finney chaired an international writing group for the African Mathematics Program that developed mathematics curricula and texts for schools in anglo-phone Africa. In 1977 he was awarded the Max Beberman Award of the Illinois Council of Teachers of Mathematics for exceptional contributions in the field of teacher education in mathematics.

performed in the right order. As these examples suggest, almost every professional field now uses mathematics of some kind.

Since 1976 the U.S. National Science Foundation has provided support for a unique multi-disciplinary response to the need for instruction in applied mathematics: the Undergraduate Mathematics Applications Project. UMAP, as the Project is called, produces lesson-length modules, case studies, and monographs from which readers can learn how to use the mathematical sciences to solve problems that arise in other fields. The applications presented by UMAP cover a broad range from chemistry, engineering and physics, to biomedical sciences, psychology, sociology, economics, policy analysis, harvesting, international relations, earth sciences, navigation, and business and vocational pursuits.

UMAP modules are self-contained, in the sense that anyone who has fulfilled the prerequisites listed inside the front covers can reasonably expect to read the modules and solve the problems without help. They cover about as much material as a teacher would put into an hour's lecture. There are exercises, model exams keyed to objectives, and answers. The modules are reviewed thoroughly by teachers as well as by professionals in the fields of application, revised, tested in classrooms throughout the world, reviewed by individual students to be sure they are as self-contained as they should be, and revised again before publication.

The modules are used for individual study, to supplement standard courses, and in combination to provide complete text coverage for courses devoted to applications of the mathematical sciences. These sciences, which I shall simply call mathematics, include probability and statistics, operations research, computer science and numerical methods as well as the elementary and advanced aspects of analysis, algebra and geometry.

UMAP case studies are not intended to be as self-contained as are the modules. The studies contain data and background information for a mathematical modeling problem as a field professional would collect it, but readers are asked to develop their own models for solving the problems. The data are real, the problems current. Teachers are given the solutions of the problems as they were originally worked out by the professional applied mathematicians who furnished the problems to the project. Each study has a teacher's guide developed through classroom use. The case studies are used in mathematical modeling courses, and may take several weeks to complete. One of their striking features is that, like the UMAP modules, they expect no previous experience with mathematical modeling on the part of either instructor or student. Nor do they require any previous knowledge of the applied field. Anyone with the right mathematical background can work through them successfully.

UMAP's expository monographs are works of eighty pages or more that make available to students in upper level courses, and to faculty in diverse fields, significant applications that are not in commercial texts. They also give users of standard texts access to additional and complementary professional methods. Like all UMAP materials, the monographs are written for students to read, and contain exercises with answers.

Although UMAP modules, case studies and monographs are similar to traditional texts in that they provide instruction for students with suitable examples and exercises, they differ dramatically in their objectives: a UMAP unit follows the logic of the practitioner, not the syllabus of a course; it presents mathematics as a natural constituent of a whole problem, not as a defined niche in a planned curriculum. Because of their allegiance to diverse masters, UMAP curriculum materials reflect both the excitement and disarray of current practice rather than the artificial order of traditional textbooks. They provide an entrée to the useful mathematics of the next decade. Here are some examples, taken from UMAP modules.

Measuring cardiac output

Brindel Horelick and Sinan Koont wrote *Measuring Cardiac Output* to teach an application of numerical integration in medicine.

Your cardiac output is the amount of blood your heart pumps in one minute. It is usually measured in liters per minute. A person awake but at rest, perhaps reading, might have a cardiac output of five or six liters a minute. A marathon runner might have a cardiac output of more than thirty liters a minute.

A change in cardiac output may be a symptom or a consequence of disease, and doctors occasionally want to measure it. One technique for doing so, one that works when the heart's output is fairly constant, calls for injecting a small amount of dye in a main vein near the heart. Five or ten milligrams will do. The dye is drawn into the heart and pumped through the lungs and into the aorta, where its concentration is measured as the blood flows past a Swan-Ganz catheter. Figure 1 shows a typical set of readings in milligrams per liter, taken every second for about twenty-five seconds.

You will notice in Figure 1 that the concentration stays at 0 for the first few seconds. It takes that long for the first of the dye to pass through the heart and lungs. The concentration then begins to rise. It reaches a peak at about 12 seconds, then declines steadily for another seven seconds. Instead of tapering to 0 at that point, however, the concentration rises slightly and holds steady. Some of the dye that went through first has begun to reappear.

The determination of the patient's cardiac output requires calculating the area under the curve that gives the concentration of the first-time-through dye. To find this curve, or at least to make a satisfactory version of it, one has to replace the real data points for the last few seconds by

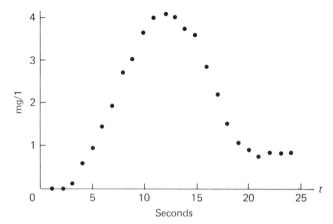

Figure 1. Typical readings of dye concentration in the aorta when 5 mg of dye are injected into a main vein near the heart at $t = 0$ seconds.

ficticious ones, as shown in Figure 2. The chosen points continue the downward trend of the points that precede them. The estimates involved in selecting the ficticious points seem reasonable, and any errors introduced by the replacement are likely to be small in comparison with other uncertainties in measurement.

The concentration curve can now be sketched, but there is no formula for it that can be integrated. This is often the case with data generated in the laboratory or collected in the field and there are standard ways to cope. On the data here there is no reason to use anything more sophisticated than Simpson's rule or the trapezoidal rule, which is precisely what Horelick and Koont proceed to do. The patient's cardiac output is then calculated by

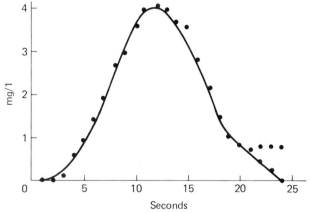

Figure 2. The curve shown here is fitted to the real and adjusted data points. Its height above the horizontal axis approximates the concentration of the injected dye passing the monitoring point in this patient's aorta for the first time.

dividing the estimate obtained for the integral (expressed in milligram minutes per liter) into the number of milligrams of dye originally injected. The result: 6.8 liters per minute.

Chemistry

Ralph Grimaldi's module, *Balancing Chemical Reactions with Matrix Methods and Computer Assistance*, shows how matrix methods may be used to balance chemical reactions. The unit gives a concrete setting for the concepts of linear independence and dependence in vector spaces of dimension four or more.

In the reaction

$$Pb(N_3)_2 + CR(MnO_4)_2 \Rightarrow CR_2O_3 + MnO_2 + Pb_3O_4 + NO,$$

which takes place in a basic solution, the atoms from lead azide and chromium permanganate combine into four other products: chromium oxide, manganese dioxide, trilead tetroxide, and nitric oxide. To find how much of each of the original reactants has to be present to produce how much of each of the products, we "balance" the reaction. That is, we find integers $u, v, w, x, y,$ and z, with the property that u molecules of lead azide plus v molecules of chromium permanganate produce exactly w molecules of chromium oxide, x molecules of manganese dioxide, y molecules of trilead tetroxide, and z molecules of nitric oxide. Schematically,

$$u\ PB(N_3)_2 + v\ CR(MnO_4)_2 \Rightarrow w\ CR_2O_3 + x\ MnO_2 + y\ PB_3O_4 + z\ NO.$$

The numbers $u, v, w, x, y,$ and z are integers chosen to make the number of atoms of each element the same on each side of the reaction. To balance the reaction, we balance the atoms.

To balance the atoms, we assign a basic unit vector to each element. It does not matter which vector we assign to which element, as long as we assign one apiece and keep track of the assignment. The assignment

$$Pb = (1, 0, 0, 0, 0)$$

$$N = (0, 1, 0, 0, 0)$$

$$Cr = (0, 0, 1, 0, 0)$$

$$Mn = (0, 0, 0, 1, 0)$$

$$O = (0, 0, 0, 0, 1)$$

will do as well as any. We use five-dimensional vectors because there are five elements.

We then replace the chemical reaction with the vector equation

$$u(1, 6, 0, 0, 0) + v(0, 0, 1, 2, 8) = w(0, 0, 2, 0, 3) + x(0, 0, 0, 1, 2)$$
$$+ y(3, 0, 0, 0, 4) + z(0, 1, 0, 0, 1).$$

You can see where the vector entries come from. For every u lead atoms in lead azide, $Pb(N_3)_2$, there are $6u$ nitrogen atoms; hence the $u(1, 6, 0, 0, 0)$ in the vector equation. For every v chromium atoms on the left side of the reaction, there are also $2v$ manganese atoms and $8v$ oxygen atoms. And so on for the other four integers, w, x, y, and z.

The idea now is to solve the vector equation for the integers $u, v, w, x, y,$ and z. To do so we rewrite the equation as a system of five linear equations in six variables. Six variables are too many for a unique solution, but we can arbitrarily assign the value 1 to the variable u to match the number of unknowns to the number of equations. We may want to change the value assigned to u later, but $u = 1$ will do for now. The resulting system in matrix form is

$$\begin{bmatrix} 0 & 0 & 0 & 3 & 0 \\ 0 & 0 & 0 & 0 & 1 \\ -1 & 2 & 0 & 0 & 0 \\ -2 & 0 & 1 & 0 & 0 \\ -8 & 3 & 2 & 4 & 1 \end{bmatrix} \begin{bmatrix} v \\ w \\ x \\ y \\ z \end{bmatrix} = \begin{bmatrix} 1 \\ 6 \\ 0 \\ 0 \\ 0 \end{bmatrix}.$$

This system of equations can be solved by a short computer program listed in Grimaldi's module. The solution given by the computer when $u = 1$ is

$$v = 2.93333, w = 1.46667, x = 5.86667, y = 0.33333, z = 6.$$

These values are not the integers we seek because they are not all integers. Once we notice that 0.03333 is about $1/30$ and 0.06667 about $2/30$, however, we know enough to scale everything by taking u equal to 30 instead of 1. The resulting solution is

$$u = 30, v = 88, w = 44, x = 176, y = 10, z = 180.$$

The module discusses what to do if at first you do not recognize the integer solution that underlies the computer's decimal solution. It also discusses an example in which reducing the number of variables to match the number of equations does not seem to work. The difficulty is traced to the fact that the reaction being balanced consists of two reactions that take place simultaneously, independently of each other. Each must be analyzed apart from the other.

Scheduling prison guards

James M. Maynard's *A Linear Programming Model for Scheduling Prison Guards* describes a linear program that Maynard developed for the Pennsylvania State Bureau of Corrections. As the newspaper clippings reproduced in Figures 3 and 4 show, the Bureau was concerned in the middle 1970's about the increasing cost of paying prison guards to work overtime. In the

At 8 State Prisons

Overtime Guard Pay Bills Keep Mounting

By The Associated Press

The Bureau of Corrections is still paying heavy overtime to keep guards on duty at the eight state prisons. Some guards are doubling their salaries through extra work.

A bureau spokesman said yesterday that in the year ended June 30 the agency paid out nearly $4 million in overtime, a boost of $750,000 over the previous year.

The bureau already had been strongly criticized for excessive overtime payments. Legislators and other officials think the state could save money by hiring more guards at regular salaries and reducing overtime payments at time and a half and double time.

Auditor Gen. Robert P. Casey, one of the critics, zeroed in yesterday on overtime at the state prison in Dallas, Luzerne County. More than $430,874 was paid during the fiscal year ended in June 1974.

Casey said 12 guards received between $10,144 and $6,707 in overtime. Eleven of the 12 guards had base salaries of $11,731. One guard has a $12,875 base salary.

"The new commissioner — William Robinson — is very aware of the problem and that, along with every other program, is being looked at very carefully ...," the correction bureau spokesman said. "He does want to cut down on the overtime."

He said former Commissioner Stewart Werner had a hiring freeze in effect because of the tight budget policy adopted by the Shapp administration.

But under Robinson, who assumed the post last month, the freeze has been lifted and 35 guard vacancies around the state are being filled, the spokesman said.

However, the overtime problem will linger.

Glen R. Jeffes, superintendent at Dallas, said vacancies alone don't govern how much overtime will be needed. Vacations and the fact that authorized staff levels are inadequate also are factors, he said.

"I have requested additional officer positions the last two years. I received no new positions..."

"Without additional officer positions I see very little impact on the reduction of overtime," he said.

Courtesy of the Associated Press.

Guards Due Windfall For Missed Breaks

From The Patriot Wire Services

About 1,700 state prison guards will be reimbursed for perhaps $1,000 each for missed coffee breaks, it was learned yesterday.

The windfall comes as a result of an arbitrator's decision earlier this month on grievances filed at eight penal institutions across the state. It may cost the commonwealth as much as $1.7 million.

Under the terms of their contract with the State Bureau of Corrections, the guards are allowed a 15-minute break every four hours.

But because of critical manpower shortages at the state's prisons, the men have not been able to take the breaks since before July, 1973.

Robert Saylor, executive director for the Bureau of Corrections, refused to comment on published reports of the reimbursement.

The guard answering the phone at the bureau's headquarters here said he had heard of the decision, but added, "We should be getting $2,000."

The arbitrator's decision was handed down on July 12, according to published reports, but the bureau made no announcement of it.

According to Jack Walsh, president of guards Local 2500 at Western State Penitentiary, the payment will be made to the guards sometime this month.

Courtesy of the Harrisburg Patriot.

Figure 3.

Senators Tour Prison Facility In Camp Hill

By MERRY BROOKS
Staff Writer

A fact-finding tour by members of the State Senate Prison Inquiry Committee yesterday at the State Correctional Institution at Camp Hill seemed more like a whirlwind campaign swing with senators shaking hands and eliciting opinions from prisoners and guards.

But in fact, the information sought by four state senators on the sixth tour of the eight state prisons yielded information that may assist the special committee in drafting prison-related legislation.

Sen. Freeman Hankins, D.-Philadelphia, committee chairman; Sen. Martin Murray, D.-Luzerne; Sen. Herbert Arlene, D.-Philadelphia, and Sen. James E. Ross, D.-Beaver-Washington, accompanied by a herd of reporters, breezed through the prison in Lower Allen Twp. in a three-hour VIP tour.

The senators engaged Ernest Patton, prison superintendent, in a give-and-take roundtable discussion before the tour began. They obtained the following information:

—The prison paid $353,000 in overtime to guards last year and expects to pay $361,000 in overtime this year. The prison needs an additional 67 guards to reduce the amount of overtime pay.

Courtesy of the Harrisburg Patriot.

Guards sought for Graterford

By The Associated Press
The head of the state Corrections Bureau says Gov. Shapp and the Legislature may be asked to provide from 80 to 100 guards at the Graterford State Prison.

The increase would raise to 400 the number of guards at the Montgomery County prison.

Corrections Commissioner Stewart Werner estimated the added guards would cost $500,000 annually.

The extra men could cut down on overtime payments to the current guards, now running about $20,000 a month.

★ ★ ★

Graterford, the largest of the state's eight correctional institutions, has about 1,600 inmates, about 200 below capacity.

Courtesy of the Associated Press.

Figure 4.

year ending June 30, 1975, for example, the Bureau paid nearly four million dollars in overtime pay, $750,000 more than it had paid for overtime work the year before. Some overtime work is to be expected, of course. It is expensive to keep a full-time staff large enough to cover peak loads, for a staff this large is likely to be underemployed much of the time. On the other hand, a staff so small that regularly scheduled guards have to work so many overtime hours that they sometimes double their salaries is also expensive, as the Bureau was finding out. Understaffing can be expensive in other ways, too, for fatigue and high inmate-to-guard ratios create dangerous tensions.

Legislators and other officials thought the State might save money by increasing the size of its regular prison staff. Maynard was hired to determine the size of the least expensive overall work force.

The goals of Maynard's investigation were to minimize the total cost of paying prison guards, while reducing the overtime work and establishing uniform work schedules in all prisons. He was able to meet the goals successfully with a linear program, the one described in his UMAP module.

Table 1 shows two work schedules for one of the Bureau's prisons, referred to here as Prison G. One schedule has parentheses, the other does not. The numbers with parentheses are the numbers of guards recommended by the linear program. The numbers without parentheses show how many guards were on duty at Prison G during the week ending September 30, 1973.

The schedules are weekly schedules divided into twenty-one periods, three shifts a day for seven days. Each box in the table shows the numbers of guards working at three different pay levels during the given shift: regular, time-and-a-half, and double time. The two numbers in the top line in each box are the numbers of guards working the shift as part of their regular weekly work schedule. The two numbers next in line are the numbers of guards working the shift at time-and-a-half. The last two numbers are the numbers of guards working at double time.

For example, Monday morning, September 24th was worked by 94 guards on regular schedules, 19 guards at time-and-a-half, 3 guards at double time. On Tuesday afternoon more than half of the 146 guards present were working overtime.

The numbers in parentheses proposed by the linear program are strikingly different from the 1973 figures. On Monday morning the model covers the work load with 117 regularly scheduled guards; where once there had been 22 overtime guards, now there are none. On Tuesday afternoon there are only 9 overtime guards where once there had been 76. The new work schedule is more equitable and less fatiguing than the old one. It is also more economical. If regular pay is calculated at $4 an hour, for instance, the new schedule for Prison G saves the State $5,216 a week.

Readers of Maynard's module are given an opportunity to follow the development of the linear program, to see the effects of various scheduling

Table 1. Data and Results from Prison G for the Week Ending September 30, 1973

Day	Shift					
	Morning		Afternoon		Night	
Monday	94	(117)	70	(131)	38	(74)
	19	(0)	61	(0)	40	(4)
	3	(0)	0	(0)	0	(0)
	116	(117)	131	(131)	78	(78)
Tuesday	94	(126)	70	(137)	36	(74)
	17	(0)	62	(9)	38	(0)
	15	(0)	14	(0)	0	(0)
	126	(126)	146	(146)	74	(74)
Wednesday	97	(116)	69	(137)	36	(74)
	19	(0)	68	(0)	27	(1)
	0	(0)	0	(0)	12	(0)
	116	(116)	137	(137)	75	(75)
Thursday	94	(128)	63	(98)	37	(74)
	41	(21)	24	(2)	34	(7)
	14	(0)	13	(0)	10	(0)
	149	(149)	100	(100)	81	(81)
Friday	74	(97)	45	(89)	37	(39)
	20	(0)	16	(0)	2	(0)
	2	(0)	0	(0)	0	(0)
	96	(97)	61	(89)	39	(39)
Saturday	57	(43)	37	(45)	26	(0)
	15	(33)	14	(6)	3	(29)
	4	(0)	0	(0)	0	(0)
	76	(76)	51	(51)	29	(29)
Sunday	53	(63)	36	(48)	25	(35)
	7	(0)	12	(0)	3	(0)
	3	(0)	0	(0)	2	(0)
	63	(63)	48	(48)	30	(30)

assumptions, and to develop a small-scale program of their own. As in the Grimaldi chemistry module, the program does not at first yield integer solutions, but by rounding the numbers of guards given by the computer to integer values and rerunning the program to determine the values of the remaining variables, one obtains a feasible solution that is close enough. It is not necessary to prove that the integer solution found this way is optimal. One can test its utility by evaluating the objective function, which gives the total amount of money paid to prison guards. If the value of the function for the integer solution is close to the value of the function for the original not-necessarily-integer solution, then the integer solution is good.

Continuous service in legislatures

Once a group of people has been elected to a legislature, the number of them who serve continuously from that time onward will normally decrease exponentially with each passing election.

The elections for the Senate of the United States are held in the fall of every even-numbered year. The senators, elected for six-year terms, take office the following January. Figure 5 shows the proportion of the 1801 Senate that remained in office after successive elections. They were all gone by 1811. The data are fitted nicely by the curve

$$y = e^{-0.029t},$$

where t is measured in months beginning in January 1801 with $t = 0$.

Thomas W. Cassteven's module, *Exponential Models of Legislative Turnover*, shows how exponential curves can be used to forecast election results, to speculate convincingly about what would have happened if a postponed election had been held on time, and to disclose suppressed data.

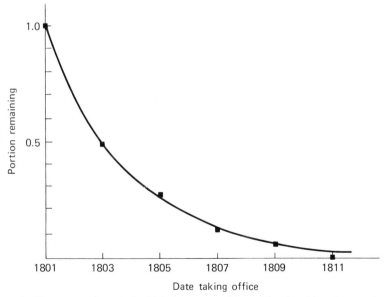

Figure 5. The proportion of the U.S. Senators taking office in 1801 who continued in office through subsequent terms. The pattern shown here, of discrete election data fitted by an exponential curve, is typical of legislative turnover. The data to be fitted may be either raw (as in Figure 6) or proportional (as in the figure above).

One of Cassteven's many interesting examples is the turnover in the membership of the Central Committee of the Communist Party of the Soviet Union. In 1957, First Secretary Nikita Khruschev, in some semi-secret infighting, succeeded in removing a number of his opponents from the Committee. Their identity was not made public, nor was their total number. Their number can be estimated, however, by a calculation based on election data from nearby years. There were elections in February 1956, October 1961, March 1966, and March 1971. From these one can calculate the exponential decay constant for the Central Committee's normal turn-over. One can then calculate how many of the February 1956 cohort should have been present after the 1961 election. It turns out that there were about 12 too few of them there. At least a dozen full members were removed in Khruschev's purge.

It is intersting to note that the decay constants for the U.S. Senate and the Central Committee of the Communist Party of the Soviet Union have been nearly equal in recent decades. For the data shown in Figure 6, the best fitting values of the decay constants are about 0.0079 (Senate) and 0.0073 (CC/CPSU). If the twelve members purged by Khruschev in 1957 are added back in, the match is even closer.

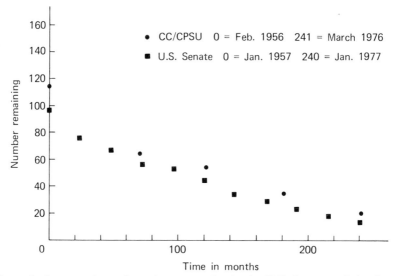

Figure 6. A comparison of continuous service in the U.S. Senate and the Central Committee of the Communist Party of the Soviet Union. The exponential decay constants of these two legislative bodies have been nearly equal in recent years. Membership in these two legislatures has been turning over at about the same rate.

Mercator's world map

Anyone who has ever wondered what the integral of the secant function is good for can find a satisfying answer in Philip Tuchinsky's UMAP module, *Mercator's World Map and the Calculus*. The unit explains how the integral of the secant determines the spacing of the lines of latitude on maps used for compass navigation.

The easiest compass course for a navigator to steer is one whose compass heading is constant. This might be a course of 45° (northeast), for example, or a course of 225° (southwest), or whatever heading is required to reach the navigator's destination without bumping things on the way. Such a course will lie along a spiral that winds around the globe toward one of the poles (Figure 7), unless the course runs due north or south or lies parallel to the equator.

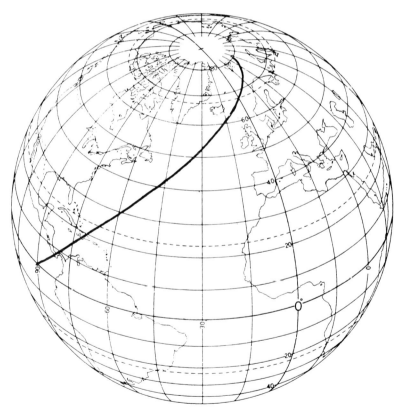

Figure 7. A flight with a constant bearing of 45° East of North from the Galapagos Islands in the Pacific to Franz Josef Land in the Arctic Ocean.

In 1569 Gerhard Kramer, a Flemish surveyor and geographer known to us by his Latinized last name, Mercator, made a world map on which all spirals of constant compass heading appeared as straight lines. This fantastic achievement met what must have been one of the most pressing navigational needs of all time. For from Mercator's map (Figure 8) a sailor could read the compass heading for a voyage between any two points from the direction of a straight line connecting them.

Figure 9 shows a modern Mercator map. If you look closely at it you will see that the vertical lines of longitude, which meet at the poles on the globe, have been spread apart to lie parallel on the map. The horizontal lines of latitude that are shown every 10° are parallel also, as they are on the globe, but they are not evenly spaced. The spacing between them increases toward the poles.

The secant function plays a role in determining the correct spacing of all these lines. The scaling factor by which horizontal distances from the globe are increased at a fixed latitude τ to spread the lines of longitude to fit on the map is precisely sec τ. There is no spread at the equator, where sec $\tau = 1$. At latitude 30° north or south, the spreading is accomplished by multiplying all horizontal distances by the factor sec 30°, which is about 1.15. At 60° the factor is sec 60° = 2. The closer to the poles the longitudes are, the more they have to be spread.

The lines of latitude are spread apart toward the poles to match the spreading of the longitudes, but the formulation of the spreading is complicated by the fact that the scaling factor sec τ increases with the latitude τ.

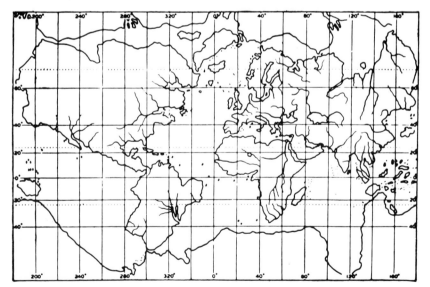

Figure 8. A sketch of Mercator's map of 1569.

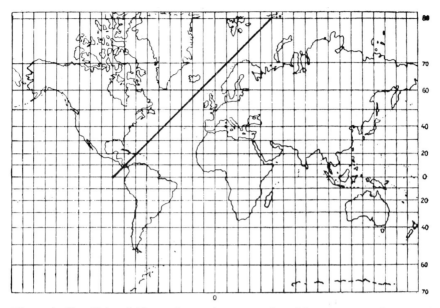

Figure 9. The flight of Figure 7 traced on a modern Mercator map. Courses of constant compass heading appear as straight line paths on a Mercator map. They are easily constructed, measured, and followed.

The factor to be used for stretching an interval of latitude is not a constant on the interval. This complication is overcome by integration. If R is the radius of the globe being modeled, then the distance D between the lines drawn on the map to show the equator and the latitude $a°$ is R times the integral of the secant from zero to a:

$$D = R \int_0^a \sec \tau \, d\tau.$$

The distance on the map between two lines of north latitude, say from $a°$ up to $b°$, is

$$D = R \int_0^b \sec \tau \, d\tau - R \int_0^a \sec \tau \, d\tau = r \int_a^b \sec \tau \, d\tau.$$

Suppose, for example, that the equatorial length of a Mercator map just matches the equator of a globe of radius 25cm. Then the spacing on the map between the equator and latitude 20° north is

$$25 \int_0^{20} \sec \tau \, d\tau \approx 9 \text{ cm},$$

whereas the spacing between latitudes 60° north and 80° north, is

$$25 \int_{60}^{80} \sec \tau \, d\tau \approx 28 \text{ cm}.$$

The vertical distance on the map between latitude 60° and latitude 80° is more than three times the vertical distance between latitude 0° and latitude 20°. The navigational properties of a Mercator map are achieved at the expense of a considerable distortion of distance.

Concluding thoughts

Mathematical reasoning penetrates scientific problems in numerous and significant ways. If the secret of technology, as C.P. Snow said, is that it is possible, then the secret of mathematical modelling is that it works. However, the process of developing and employing a mathematical model is both more subtle and more complex than is the traditional solution of mathematics textbook problems. Real models frequently have to be constructed in the presence of more data than can be taken into account; their conclusions are often drawn from calculations in which good approximations play a greater role than do exact solutions; very often there are conflicting standards by which solutions can be judged, so whatever answers emerge can only rarely be labelled as right or wrong. Students using UMAP modules, case studies, or monographs experience mathematics in its scientific context, and leave the classroom better equipped to face real demands of mathematical modelling in business, research, and government work.

The Decline of Calculus—
The Rise of Discrete Mathematics

Anthony Ralston

> Analysis is the technically most successful and best-elaborated part of mathematics.
>
> John von Neumann [1951]

> There is a simple and basic fact about a computer which will, in the decades and centuries to come, affect not so much what is known in mathematics as what is thought important in it. This is its finiteness.
>
> Wallace Givens [1966]

Calculus is one of the great triumphs of the human intellect. For this reason alone no educated person should be without some knowledge of it. When, in addition, you consider the panoply of intellectual and practical conquests of classical analysis, whose foundation is calculus, it is small wonder that calculus has been for so long the basis of all college mathematics study. It may well surprise the reader then that the purpose of this essay is to argue that the position of calculus in the college mathematics curriculum is ripe for change and, to a degree, decline.

If you believe—as I do—that mathematical research requires as high a degree of creativity and imagination as any form of intellectual activity, then it follows that mathematicians—good ones anyhow—must be the least conservative people around as they ply their trade, where I use "conservative" in the sense of sticking to established modes of thought. Indeed, the successful research mathematician must be a radical, ready to embrace—or, at least, to consider—the wildest of new ideas. But not only in their direct professional activities are mathematicians radicals. Many are educational radicals also, at least insofar as they view primary and secondary education. It is not my purpose here to discuss the "new math" but, whatever you may think about it, the ideas were radical. (Some people

Anthony Ralston is a Professor in the Department of Computer Science at the State University of New York at Buffalo, a department of which he was chairman from 1967 to 1980. He was trained as a mathematician with an S.B. (1952) and Ph.D. (1956) from M.I.T. He has directed major computing centers at SUNY at Buffalo (1965-70) and the Stevens Institute of Technology (1960-65), and was President of the Association for Computing Machinery (1972-74) and of the American Federation of Information Processing Societies (1975-76). He has authored or edited eight books on various topics in mathematics and computer science.

would say "reactionary" but that is only another proof that the political spectrum is really a circle.) In any case, how strange it is that research mathematicians are the most conservative types around when it comes to their own turf, the college and university mathematics curriculum.

At some slight risk of oversimplification, it may be said that the only significant change in this century in the basic college mathematics curriculum has been that college algebra and trigonometry, which were the subject matter of the standard first year college mathematics course until after World War II, have descended into the high school curriculum to be replaced by two years of calculus and analytic geometry and then by calculus and linear algebra. Until the Second World War calculus was usually not reached by college mathematics students until the sophomore year. In the 1960's there were some short-lived experiments, notably at Dartmouth, to include a semester of "finite mathematics" in the first two-year sequence but in almost all cases this course has now been replaced by linear algebra. Recently college algebra and trigonometry have been making a comeback in college mathematics as the mathematical preparation of entering freshmen has seemed to decline. But the basic point is unaffected by these variants; there has been little variation in the first two years of college mathematics this century.

I would not argue that things should have been much different. Seldom should there be rapid changes or discontinuities in the presentation of the subject matter of any discipline at a primary, secondary or beginning university level. Fundamental changes in knowledge are rare; more rarely still do such changes imply commensurate changes at basic educational levels. Moreover, we still understand so little of what distinguishes good educational approaches from bad, and have so little ability to measure usefully anything we do in education, that sharp changes should be viewed with great skepticism and should be undertaken only for the most compelling reasons.

Still, I suggest the need for such a revolution. Its cause? The invention and development over the past three decades of the digital computer, perhaps the most important development in science and technology since the invention of the printing press. In any case, it is a development which will not only have profound affects on human life and the social fabric, but which will also—and this is the point here—have a most important influence on the problems on which scientists work and, in particular, on the mathematics they use. (Which is not to say, I emphasize here, that calculus and classical analysis will not continue to enjoy much success; it is only to say that their position of dominance in mathematics and its applications is about to be challenged.)

What are—or should be—the contents of this revolution? Nothing less than some kind of curricular equality (at least) for discrete analysis and classical continuous analysis. By discrete analysis I mean to include those branches of mathematics which focus mainly or entirely on discrete objects including combinatorics, graph theory, abstract algebra, linear algebra, number theory and discrete probability. This equality should manifest itself at least by the provision of a discrete mathematics-based, freshman-sophomore sequence as an alternative to the standard calculus sequence. (After such a sequence, the standard third year course, for mathematics majors anyhow, would be calculus.)

> *... for many who deal directly with computers . . . the mathematics most important to them tends not to be calculus, but areas of discrete mathematics.*

The essential motivation for this proposal is that for many who deal directly with computers, including at least almost all computer scientists and social, behavioral and management scientists, the mathematics most important to them tends not to be calculus, but areas of discrete mathematics. Beyond this, however, I believe that the trends in mathematical research itself are increasing strongly in the direction of discrete mathematics, a claim which can be supported, although surely not proved, by studying trends in the publication of research mathematics. Aside from any empirical evidence, however, this proposition is bolstered by the observation that, to a considerable degree, the wellsprings of mathematics have always been in the applications of mathematics. Today it is computers generally and computer scientists in particular which generate the need for applications of mathematics in greater volume—and at a much more rapid rate of increase —than does any other area of science or technology. Since the mathematical problems generated by computers and by computer scientists over-whelmingly require discrete rather than continuous mathematical tools, it is hardly surprising that research in discrete mathematics is rapidly increasing. In the remainder of this essay I shall try to give some insights into why discrete mathematics plays such an important role in computer-related mathematics.

First, let me note what is, at least at present, a rather unimportant reason for focusing on discrete mathematics. The wonderful successes of classical analysis have resulted despite the fact that the physical situations they model are actually discrete systems. For example, the motion of a body is, in fact, the aggregate of the motions of its single, discrete molecules. The number of these molecules is, however, so vast that only by considering them as an aggregate has it been possible to derive meaningful results. And, of course, the approximation made in so doing is so close to reality that the results derived from it are generally very accurate indeed. Somewhat ironically, the actual calculations done on computers using the formulas of classical analysis can only be performed by first discretizing these formulas by, for example, replacing derivatives by differences and integrals by sums. (This discretization, however, results in a much less detailed model than would be the case if individual molecules themselves were considered.) We have, therefore, a situation in which essentially discrete phenomena are modelled by continuous functions which must then be discretized for calculational purposes.

Would it not make more sense to treat the discrete problem discretely in the first place? For the present, only theoretically. The enormity of the calculations required to do this are still well beyond the capacity of the largest computers available. But this may not always be true, as the advent of microprocessors spurs the development of vast networks of computers. If and when such calculations become possible, they may provide a connection between physical reality and calculational practice which will be most helpful in explaining and using the underlying physical principles. But this age has not yet arrived: the analysis of physical reality is not now a compelling reason for stressing the importance of discrete analysis.

Algorithms

The application of discrete analysis on which I wish to focus is its use in the analysis of algorithms, well-defined procedures for calculating some numeric or symbolic quantity. Since almost all computer programs are implementations of algorithms, it is extremely important to be able to answer two basic questions:

1. Is a procedure an algorithm? That is, does it unfailingly lead to the desired result or, at worst, to an indication that it cannot solve the problem?
2. How effective (i.e., how efficient) is a particular algorithm (either absolutely or in relation to other possible algorithms for the same problem)?

Finding answers to these questions is the subject matter of the *analysis of algorithms*.

To illustrate the issues and kinds of mathematics involved in the analysis of algorithms, let me consider a particular problem (which is not itself of much intrinsic importance): Each room in the Baltimore Hilton Inn has, instead of a standard lock, a dial on which to enter a four digit combination. So long as the correct four digit combination appears in any sequence of digits that are dialed, the lock will open. Suppose that a burglar wishes to break into a particular room and, having no information about the combination, naturally wishes to dial as few digits as possible. Although as many as 40,000 digits may be required (since there are 10^4 possible combinations, each requiring 4 digits), proper planning would allow fewer digits since, for example, having dialed *abcd*, dialing *e* effectively tests the combination *bcde*. The Baltimore Hilton problem is to determine the minimum number of digits that must be dialed in order to check all possible combinations, and to find an effective algorithm for generating a sequence with this minimum length. (Readers familiar with graph theory will recognize this problem as just a restatement of the de Bruijn cycle problem.)

(We note, by the way, the unfortunate terminology used with locks. In mathematical terms what we are seeking is a method to generate all *permutations*, not combinations. What is commonly called a combination lock should really be called a permutation lock.)

We begin, as mathematicians are wont to do, by generalizing the problem. Let us assume that there are m distinct symbols (in the Baltimore Hilton problem, $m = 10$) and that we wish to generate a string of minimum length which contains all combinations—that is, all permutations—of these symbols of length n ($n = 4$ above). A lower bound on the length is $m^n + n - 1$ since there are m^n possible combinations, the first one can be generated with a string of length n and each additional symbol could generate at most one of the other $m^n - 1$ combinations.

Is this lower bound achievable or must the minimum length sequence contain some repetitions of sequences of length n? The answer follows from a well known theorem in graph theory, a key branch of discrete mathematics. This theorem shows that, in fact, all repetitions can be avoided so that the minimum length we seek is precisely the lower bound.

To see this, we consider a graph with m^{n-1} nodes in which each node is labeled with one of the m^{n-1} combinations (i.e., permutations) of our m symbols of length $n - 1$. Figure 1 shows a portion of such a graph in which we have assumed that our m symbols are the digits $0, 1, \ldots, m - 1$. The node shown with label $m_1 m_2 \ldots m_{n-1}$ has m edges emanating from it to

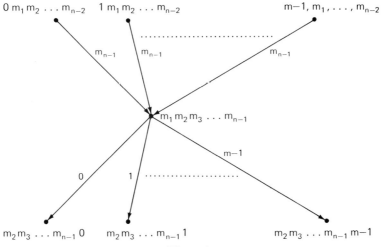

Figure 1.

nodes labeled $m_2 m_3 \ldots m_{n-1} i$ with $i = 0, 1, \ldots, m - 1$; each edge is labeled with the corresponding i. Moreover, each node has m edges incident on it, each labeled m_{n-1}, from nodes labeled $im_1 m_2, \ldots, m_{n-2}$ again with $i = 0, 1, \ldots, m - 1$.

The theorem required for our analysis states that a graph such as that suggested by Figure 1 (technically, a connected, directed graph) has a path including every edge once and only once (technically, an *Eulerian path*) starting at *any* node if each node has the same number of edges incident on it as emanating from it. (This is merely a formal solution to the popular child's pastime of tracing over a geometric figure without ever lifting the pencil from the paper or going over the same segment twice.) Each node in the graph represented by Figure 1 has m edges entering and leaving it, and it is connected, so it must have an Eulerian path beginning at any node.

Suppose now we start with the string $m_1 m_2 \ldots m_{n-1}$ and proceed from the node with that label on an Eulerian path around the graph. Each time we traverse an edge we add the label of that edge to our string. (Thus, when going from node $m_1 m_2 \ldots m_{n-1}$ to node $m_2 m_3 \ldots m_{n-2} i$ we add i to the

---

We've proved that a solution exists and have a way of constructing that solution. But a computer scientist, not satisfied yet, would want to know if the algorithm could be efficiently implemented.

---

string.) As each edge is traversed, the last n symbols on the string form one of our combinations. Moreover, because of the labeling of the nodes and edges, no combination can appear more than once. Since there are m^n edges (m^{n-1} nodes and m edges emanating from each node), when we complete our Eulerian path our string will have precisely $m^n + n - 1$ symbols on it containing, as advertised, the m^n permutations of length n (i.e., lock combinations) without repetition.

Using this proof of the existence of a string of length $m^n + n - 1$ we can easily formulate an algorithm to generate this string:

1. Start with a string of n 0's. (This corresponds to starting at the node labeled with $n - 1$ zeros and traversing the edge labeled 0 which is a loop which takes you right back to the node you started from.)
2. At each stage add a nonzero symbol to the string if this does not give a permutation already generated; otherwise add a 0.

Although not quite trivial, it is not too hard to prove that this process cannot terminate until you return to the initial node for the $(m + 1)$st time and that this will not happen until all edges have been traversed. (An Eulerian path which ends up where it started is called an *Eulerian cycle*.)

Is this the end of the problem? Perhaps, as a mathematician would view it. We've proved that a solution exists and have a way of constructing that solution. But a computer scientist, not satisfied yet, would want to know if the algorithm could be efficiently implemented on a computer. The biggest problem is in step 2, where each potential symbol must survive an extensive "repetition check" to insure that adding it will not duplicate a combination that had previously been created. For the Baltimore Hilton problem with $m = 10$ and $n = 4$, $10,000$ words of memory are required to store the $10,000$ possible combinations so that each can somehow be marked when it appears in the sequence. Moreover, we may have to look in this table of $10,000$ values up to nine times each time we wish to add a symbol until we find a sequence which has not appeared. In general the table would have to be m^n words long with up to $m - 1$ table lookups each time.

One possible way to improve the space efficiency (or, as it is more often called, the *space complexity*) would be to store a sequence of $10,000$ bits (binary digits) instead of words, marked 0 if the corresponding sequence has not appeared and 1 if it has. But this would be at the expense of the *time complexity* since looking something up in a table of bits is typically quite a bit less efficient than looking up information in a table of words.

What we would like is a constructive algorithm which enables us to add a symbol to the sequence directly without requiring a table search. Such algorithms do exist, but they are rather complicated and so we only sketch one here.

1. Begin with the single digit $m - 1$.
2. Then append the $m - 1$ strings of length n starting with $(m - 1)$ $\cdot (m - 1) \ldots (m - 1)(m - 2)$, and where each subsequent string has all the symbols the same except the last which is reduced by one each time. (In the Baltimore-Hilton case we would have, after step 2,

$$9\ 9998\ 9997\ 9996\ \ldots\ 9990.)$$

3. Next append the $(m - 1)^2$ strings of length n starting with $(m - 1)$ $(m - 1) \ldots (m - 1)(m - 2)(m - 2)$, and where each subsequent string has $n - 2$ $(m - 1)$'s followed by the two symbols from among $0, 1, \ldots,$ $m - 2$ which form the largest two digit number less than that in the previous string. (In the Baltimore-Hilton case, this yields

$$9988\ 9987\ \ldots\ 9980\ 9978\ \ldots\ 9970\ \ldots\ 9900.)$$

4. Now we must be necessarily vague because of the complexity which arises. We continue with strings of length n, systematically introducing fewer and differently placed $(m - 1)$'s at each stage. Each string of length n does, however, depend upon its predecessor in a well-defined way.
5. Finally we append the corresponding string for the case $m - 1$ and n, which we generate by applying steps 1 through 4 with m replaced by $m - 1$; then we apply this step with $m - 1$ replaced by $m - 2$, etc. Since

the string for $m = 1$ and any n is just n zeros, this process ends after $m - 1$ applications of steps 1 through 4.

There are several things worth noting about this algorithm. First, since each string of length n depends upon its predecessor in a well-defined way, the algorithm does not require a memory of just which strings have been generated previously. And because the rule by which one string is generated from the next is rather simple, the algorithm is efficient from the time as well as space complexity point of view.

Steps 2 through 4 involve an *iterative process* (i.e., generating one string after another) which is characteristic of many problems in discrete mathematics and algorithmics. Step 5 involves a *recursion*, that is a solution of a problem for a particular parameter (m) which depends upon successively smaller values of that parameter ($m - 1, m - 2, \ldots, 1$). Recursion is one of the most powerful tools available to both the discrete mathematician and computer scientist.

Finally, although our discussion of the algorithm did not display this, it turns out, rather surprisingly perhaps, that the details of the algorithm are quite a bit simpler when n is prime than when it is not.

To summarize, this problem, together with its associated algorithms, typifies the issues involved in the analysis of algorithms and the interaction of various areas of discrete mathematics (graph theory, combinatorics, even number theory) with the analysis and design of algorithms. One example, however, doesn't prove a case. But my belief is that the range of problems associated with the design and analysis of algorithms is so broad, is growing in importance so rapidly, and will remain important for so long, that research in discrete mathematics will progress very rapidly to provide the results and techniques needed in algorithmics. Thus, I come back to the point made at the beginning of this essay: it is time—if not past time—for discrete mathematics to be represented in the undergraduate mathematics curriculum as at least an equal partner with classical analysis.

References

[1]   Daniel Greenspan, *Discrete Models*, Addison-Wesley, 1973. (The best reference on the desirability of discretizing the basic laws of mathematical physics.)
[2]   Donald E. Knuth. *The Art of Computer Programming*, V. 1, 2, 3, Addison-Wesley, 1968, 1969, 1973. (The classical references on algorithms, their analysis and the mathematics needed for this.)
[3]   Anthony Ralston, "Computer Science, Mathematics and the Undergraduate Curricula in Both," *American Mathematical Monthly* (to appear). (A long paper which expands on many of the ideas in this essay.)
[4]   Anthony Ralston, "A New Memoryless Algorithm for De Bruijn Sequences," *Journal of Algorithms* (to appear). (In which the algorithm sketched above is considered in detail.)

Mathematical Software:
How to Sell Mathematics

Paul T. Boggs

When faced⋅ with the question, "What does a mathematician do if he doesn't teach?," most people have one of three answers: "I don't know." "He solves problems." Or, "He analyzes data." The average person does not think in terms of mathematical modeling or qualitative analysis, much less in terms of abstract theories. This, coupled with the prevailing view that university mathematics is strictly theoretical (ivory tower), and hence of no practical value, sometimes makes the justification of support for research in mathematics a trying undertaking. Outside of the National Science Foundation, managers in industry and government do not always respond well to vague arguments of "eventual" applicability.

Those of us in the "mission-oriented" research support agencies are often called upon to defend our programs and to justify increased (or even level) funding. While other subjects are almost as poorly understood as mathematics, we seem to be at a special disadvantage in that we have no tangible results to display as the product of our efforts. "Increased understanding" and "insight" do not, unfortunately, always compete well with lasers, computers on a chip, or some fantastic new material. That certain products or devices would have been impossible without a deep mathematical understanding is often a very difficult argument to make convincingly.

While mathematics will probably never have actual products to display,

Paul T. Boggs is Associate Director of the Mathematics Division of the U.S. Army Research Office and Director of the program in Numerical Analysis and Computing. He received a B.S. in mathematics from the University of Akron in 1966 and a Ph.D. in computer science from Cornell University in 1970. He has taught computer science and mathematics at the University of Kansas, at Rensselaer Polytechnic Institute, at the University of North Carolina, North Carolina State University, and Duke University. His active research interests include numerical methods for nonlinear optimization and for nonlinear systems of equations. He is a member of the Society for Industrial and Applied Mathematics and the Association for Computing Machinery where he was a member of the Board of Directors of the Special Interest Group in Numerical Mathematics.

mathematical software can almost serve the same purpose. When people think of a mathematician solving problems, they often have in mind some sort of computer solution. Since mathematical software is an implementation on a computer of an algorithm for solving a mathematical problem, it is both concrete and easy to identify. In addition, mathematical software can and often does incorporate to varying degrees much research in both applied and pure mathematics. Therefore it provides a convenient vehicle for explaining an important value of mathematics and for justifying further research in many fields. Furthermore, the relatively recent recognition of the difficult and often subtle problems encountered in producing such software has given rise to an increased acceptance of mathematical software as a legitimate research field in its own right.

Mathematical software, however, can be both a blessing and a curse. What makes it so valuable is its inherent ability to achieve "technology transfer" quickly and effectively. In this sense, mathematical software can be regarded as a natural end product of mathematics. For example, if a more effective procedure is discovered for solving a certain class of problems, this procedure may be incorporated into a high quality piece of mathematical software and used almost immediately by anyone desiring to solve such a problem. More importantly, the user need not be concerned with the details of the algorithm or with the underlying theories concerning the technique. Of course this is a simplified scenario—the road is seldom either so smooth or direct—but the potential for rapid transfer from the researcher to the user is real and very important. The dangers arise when software is oversold and managers and users begin to think that all of their problems are solved.

The importance of mathematical software can be underscored by the burgeoning microprocessor industry which will make computers vastly more widespread than they already are. Such devices are now in use or suggested for use in applications ranging from health care systems to nuclear reactors, from automobiles and airplanes to military systems. In most such applications they are intended to monitor and control processes, a use which often involves significant numerical computation. One hardly needs to mention the potential dangers of poor software controlling a nuclear reactor.

I shall attempt in this essay to describe mathematical software and to discuss the advantages and dangers of using it as a device to justify research support in mathematics. I hasten to add that no one should infer from this discussion that all the people with whom we deal are incapable of appreciating mathematics. Many managers must be concerned with a variety of pressures, including short-term return on investment. Some areas of mathematics (and physics, chemistry, etc.) are therefore simply more palatable to some audiences than to others. In any event, other mathematicians should understand the need to have the profession justified on

occasion and should be grateful for an area which can be readily used to defend many fields.

What is mathematical software?

Mathematical software is usually distinguished from data processing, or systems, software in that the latter does not depend to any significant degree on "sophisticated" mathematics, but rather on the manipulation of data. Despite its sophistication, mathematical software is generally regarded as less important than data processing software primarily because of three widely held misconceptions. The first is that the overwhelming majority of programs is concerned with data processing or systems. The second is that the increasing speed and capacity of modern computers make further advances irrelevant, and the third is that mathematical software is easy to produce. An often heard statement in defense of the first misconception is that ninety percent of all programs are in COBOL, a business-oriented programming language. An understanding of the fallacies of the other two led to the quip, "Yes, but ninety percent of the content is in FORTRAN," the major language for mathematical software.

For the first misconception, it suffices to consider the facts. J.R. Rice [1] has collected data in an attempt to clarify the division between scientific and non-scientific computing. His analysis indicates that, contrary to popular belief, about fifty percent of the total computing in the United States (and possibly as high as seventy-five percent in the Department of Defense) is devoted to scientific activities. Translated into dollars this means that between one and a half and two *billion* dollars annually are being invested by the federal government alone in the development and use of mathematical software.

The second misconception is essentially a statement that improvements in computer hardware alone have led to our increased ability to solve problems. To illustrate the progress that has been made in the development of methods, recall that the speed of the early computing machines was of the order of thousands of arithmetic operations per second. Today, the fastest machines—the array, pipeline and vector processors—operate at a top speed of the order of hundreds of millions of operations per second— an increase of about five orders of magnitude. During that same period, however, advances made in algorithms for some important classes of problems have been far more remarkable. Again J.R. Rice reports (in [2]) that for some nonlinear partial differential equations, algorithms of 1975 were about ten orders of magnitude faster than those of 1945! While not all advances have been so dramatic, it is clear that algorithmic improvements have enabled the solution of far more problems than have increases in hardware speeds. When coupled together, the progress has been truly impressive, but each success invites the use of more sophisticated mathematical models which continue to test the ingenuity and skills of both the mathematicians and computer engineers. Indeed, there are current problems—for example, in numerical weather prediction—which by today's methods would totally exhaust the capacities of any existing or even contemplated computer system.

The final misconception—that mathematical software is easy to produce —is more complicated. It often stems from elementary computer science texts in which mathematical software is illustrated by a simple code to solve a quadratic equation. This presentation leads to the impression that mathematical software "only" involves the rather routine implementation of a few formulas. That there can be many deep and subtle complications is a fact which is not often clearly elucidated. For example, a straightforward implementation of the usual formula for solving a quadratic equation can produce totally inaccurate results (see, e.g., Hamming [3]). From this and similar examples have grown the realization of the need for detailed analysis and careful implementation of numerical procedures.

To appreciate the difficulties associated with mathematical software and the necessity for concerted efforts to produce high quality products, a few words about the underlying algorithms and the field of numerical analysis are in order. Although the distinction is fuzzy, numerical analysis concentrates on the development and analysis of the basic algorithms, while mathematical software is concerned with the analysis of the details of implementation and testing.

Modern numerical analysis began with the development of high-speed digital computers in the late forties. Prior to that there was a collection of carefully designed procedures for hand calculation, but the possibility of large scale computations was, of course, not generally considered. One of the major difficulties here is the imposition of finite-precision arithmetic which gives rise to the well known, but again widely misunderstood,

---

The speed with which some computations can be rendered useless by the cumulative effect of small errors is quite amazing.

---

phenomenon of round-off error. Many people erroneously believe that, simply because the computer uses fifteen significant digits, their answers will be accurate to fifteen digits. However, the speed with which some computations can be rendered useless by the cumulative effect of small errors is quite amazing.

One of the major successes of numerical analysis has been the discovery and analysis of certain procedures which are not catastrophically affected by round-off error. Another major insight has been the recognition that certain problems are inherently computationally difficult—the so-called "ill-conditioned" problems.

As an example of ill-conditioning consider the two linear equations in the two unknowns x and y

$$ax + by = c$$
$$dx + ey = f$$

where $a, b, \ldots, f$ are given. The problem is to determine x and y. The numbers $a, b, \ldots, f$ make up the problem data. It happens that for some data the equations can be solved almost as accurately as the data are given, e.g., if the data have three significant figures, x and y can be determined to three figures. Another way of putting this is that changes, or perturbations, in the fourth significant digit of the data only induce changes in the fourth digit of the answer. For other data accurate to three places, however, changes in the fourth digit of the data could cause changes in the *first* digit of the answer thus rendering any computation useless. It is very important to note that this is a property of the problem (data) and not a property of the algorithm chosen to solve this problem. However, almost any method will introduce some round-off errors into the data. A good algorithm therefore will attempt to minimize this effect in order to achieve as wide a domain of problems as possible for which "reasonable" answers will be computed.

It is unfortunate that much mathematical software has been developed with a lack of awareness of some of these issues. Complicating this picture is the fact that many complex programs are critically dependent on certain parameters that are rarely discussed and poorly understood. Thus in practice, different implementations of the "same" algorithm often perform vastly differently because of the complex interaction of these factors. Even today, many computer center libraries contain numerous algorithms of widely varying quality for solving a variety of mathematical and statistical problems. There is much duplication and many a program can only be used with confidence by the author since no reasonable description and documentation exist. A potential user is thus confronted with an often bewildering hodge-podge of codes with virtually no professional advice. It is against this background that many people have finally come to realize the importance of high quality mathematical software and the dangers of poor quality software.

It is not possible to give a universally accepted complete description what constitutes high quality mathematical software. Nevertheless there a few characteristics which are fairly well agreed upon. As the ab example illustrates, a basic requirement of such mathematical softwa that the computation should continue only as long as meaningful r can be obtained. This implies that estimates must be derived which in when the boundaries of the problem domain are exceeded. If it is mined that further computation is fruitless, the software should stop clear indication of what the difficulty is. (There is little as frustrat user as getting an incomprehensible error message as the only out

someone else's program.) It is safe to say that, as of today, there ᴀ few problem classes for which such estimates have been derive, implemented in widely available routines.

Related to this is the observation that, as the problem domain bou ries are neared, the performance of most algorithms deteriorates. It therefore also desirable that an algorithm should degrade gracefully ᴀ these boundaries are approached. In some cases, e.g., in certain iterative procedures, it may be possible to detect slow progress and to so inform the user. It may also be possible to provide an estimate of the cost to obtain the solution to within the desired accuracy.

There is little as frustrating to a user as getting an incomprehensible error message as the only output from someone else's program.

Certainly any high quality product must adhere to the highest standards of the general software industry. These include standards of documentation, style, testing and portability. The issue of portability (the ability to run the same program on many machines) is especially important since some machine- and system-dependent features need to be explicitly taken into account. For example, it makes no sense to require convergence to twenty significant digits on a machine which carries only ten. Portability is definitely not achieved by merely writing the program in standard FORTRAN.

Additionally, issues of human engineering are beginning to receive attention. The ease of use, the ability of the user to interact with the algorithm, the interface with symbolic mathematical processors, and the clarity of the output (including its suitability for graphical display) are all difficult topics in need of considerable research.

One does not normally write a piece of software of this quality in isolation. Often the program is to be part of a larger set of routines for some special purpose or even part of a library of routines for varied purposes. In such cases the programs must be considered as part of this larger entity and issues of consistency and compatibility must be addressed. Such questions add a dimension which until recently has received rather little attention.

Currently there are only a few problem classes so well understood that investment in such software is warranted. Examples include linear systems of equations, definite integration, ordinary differential equations, eigenvalues of symmetric matrices, and several others. Again this is not to say that all linear systems or all ordinary differential equations can be solved since there are still many open questions concerning procedures for very large linear systems and so-called "stiff" differential equations. What is meant,

and what should be clear from the above, is that for well-conditioned instances of such problems, known algorithms perform acceptably well. For many other classes of problems, however, no consensus on the best algorithm exists and hence a costly investment in software is premature. In these areas there continue to be extensive efforts in developing, testing and analyzing various approaches. Experimental software is of course written in support of these activities and some of it has been distributed and used successfully.

A primary force behind the development of much mathematical software has been the federal government. A very early example is NASTRAN, a code to analyze mechanical structures, produced by the National Aeronautics and Space Administration. Although limited in both its mathematical scope and its implementation, it continues to be an effective tool for many engineers in certain situations. In another area, the National Institute of Health has sponsored the development of a general statistical analysis package called BMD. This set of routines has been very widely distributed, but has met with mixed success over the years. More recently, Argonne National Laboratory has embarked on a program to produce very high quality packages in several areas. The Argonne projects include EISPACK for matrix eigenvalue problems, LINPACK for linear equations and related tasks, MINPACK for function minimization, FUN-PACK for special functions, and several others. Most of this work has been done by a combination of Argonne personnel with university and industrial experts. Many other projects are being undertaken by the Department of Defense, Department of Energy, National Science Foundation and National Bureau of Standards.

Private industry has also invested heavily in mathematical software but most of their work is proprietary. There are however a few private companies marketing general purpose mathematical software. Probably the two best-known are the International Mathematical and Statistical Libraries and the Numerical Algorithms Group. Both of these provide extensive libraries of routines for a wide range of problems. Their products are widely distributed and have been successfully used on many computer systems. The impact of their work has been quite substantial in demonstrating the effectiveness of good software at very reasonable prices.

University scientists have been very active in many aspects of mathematical software, but, owing to the high cost of actual development, have not produced much software of production quality without substantial outside support.

From this description of mathematical software, one should not be surprised to learn that such products are expensive. Estimates vary according to the application, but often range from fifteen to fifty dollars per line of final code.

The dangers

As I mentioned earlier, mathematical software provides one of the best means of technology transfer. With a well-written code even non-experts can easily use the latest algorithm for solving their problems. They need not be concerned with the details but can proceed directly to its use. The most obvious examples are the elementary functions such as the square root and the trigonometric functions. Most people neither know nor care how these functions are approximated. (Even many of those who think they know how it's done probably do not.) But they use them freely to build more complicated programs, which is as it should be—progress being made by building on the work of others. The aim of the software community is to create a set of increasingly more powerful and sophisticated tools which can be reliably used by experts and non-experts alike.

Even at the high cost of producing high quality software, the potential for savings makes its production very tempting. A routine which is used thousands of times by scientists and engineers across the country can more than pay for the research as well as the developmental costs. Even a seldom used routine which produces a correct answer in a difficult but critical situation can be of inestimable value.

On the dark side there is the danger, as with all powerful tools, of misuse. Selection of the wrong routine can result in erroneous answers or much poorer performance than necessary. A routine can be applied outside of its range in a way which is impossible to detect. For example, much of classical statistical analysis is based on the assumption of an underlying normal distribution. If this assumption is not warranted, the numbers produced will be meaningless. Unfortunately, there are still those who believe that if numbers are produced by the computer, they must be right.

A critical problem is that managers are sometimes deceived into thinking that with a computer and a good library of programs, they no longer need a mathematician or statistician on their staff. This is a case of over-selling, or of failing to provide an honest assessment of the limitations of the product.

With such tools readily available there is a tendency among some to rush to the computer without doing any preliminary analysis or critical thinking about their problem. In some cases, such a practice is institutionalized in the sense that certain computer analyses are required even though the results provide little or no (or even misleading) information. People in this situation often prefer poor codes which always return "answers" to those which warn when problems are present. Of course this is not the fault of mathematical software, but the mere existence of such tools encourages this type of mindless activity.

On balance, it seems clear that the success of mathematical software in several areas and the growing demand for more high quality products overshadow the potential for harm from poor or misused codes. Furthermore, current research trends should give rise to even better methods and safer software. There is no doubt that problems are solved today which were considered intractable ten years ago and that quality mathematical software has been an important vehicle in making new methods widely available. Thus it is clear that further investments in research in mathematical software, numerical analysis and many other areas of mathematics will continue to have high payoff over a broad scientific range.

References

[1] J.R. Rice. "Software for Numerical Computation." In *Research Directions in Software Technology* (Ed., Peter Wegner). MIT Press, 1979, pp. 688-708.

[2] ———. "Algorithmic Progress in Solving Partial Differential Equations." *ACM SIGNUM Newsletter* 11 (December 1976).

[3] R.W. Hamming. *Calculus and the Computer Revolution*. CUPM Monograph, Mathematics Association of America, 1966.

Physics and Mathematics

Hartley Rogers, Jr.

Mathematics and physics are closely related as disciplines. Their histories are intertwined, sometimes in the person of a single figure such as Newton. Each has helped to give form and emphasis to the other. The practice of physics requires a broad knowledge of mathematics, and the mathematician who seeks wider understanding in his or her own field does well to become familiar with the classical areas of physics.

The two disciplines are also closely interrelated in the educational introduction that is given to each at the beginning college level. The physics teacher depends upon the mathematics for (initially) a systematic coverage of vectors, calculus, linear algebra, and differential equations. The mathematics teacher seeks to respond in a helpful way, in part because more of the students in introductory mathematics will be going on to concentrate in physics or in fields like engineering that depend on physics than will be going on to concentrate in mathematics. The central place of calculus in introductory mathematics courses is evidence of this need and this response.

The relationship is, however, not a fully happy one. The physicist feels pressed to make as much use as possible, as soon as possible, of the mathematical ideas upon which his subject depends. He frets that the mathematics teacher, striving for a fullness of coverage, is lingering fondly

Hartley Rogers, Jr. is currently a Visiting Scholar in Philosophy at Harvard University. He received his undergraduate education at Yale University (B.A. in 1946) and did graduate work in physics and mathematics at Cambridge, Yale, and Princeton Universities (Ph.D. in 1952). After three years as a research instructor at Harvard, he moved to the mathematics department at the Massachusetts Institute of Technology, where his research centered in mathematical logic and related areas. He is the author of *The Theory of Recursive Functions and Effective Computability* (McGraw Hill, 1967). From 1971 to 1973 he was Chairman of the Faculty at M.I.T., and from 1974 to 1980 served as Associate Provost. He has a particular interest in the intuitive logic and conceptual content of the natural sciences and of mathematics. He is currently working, as well, on problems in applied mechanics and applied probability.

and unnecessarily on details of rigor that, if anything, obscure the simplicity, usefulness, and intuitive naturalness of mathematical concepts. The mathematician, on the other hand, argues that a quick and informal treatment will prevent the student from developing an adequate sense of the independent conceptual reality of mathematical models. If mathematics remains merely a set of rules-of-thumb and algorithms to be memorized and to be applied where appropriate, the student will not develop the ability to create, adapt, or modify mathematical models, and the educational purposes of the physics teacher become more difficult to realize.

The resolution of this unhappiness is not a simple matter and will not be attempted here. The pedagogical problems at issue are more than a difference in emphasis between two disciplines. They raise questions of modes of understanding, of cognitive psychology, and of epistemology, that are difficult even to formulate. This does not mean that our current teaching, with its interdisciplinary tensions, is not effective. Indeed, the similarity of approach and of independently generated course materials from college to college is evidence of an empirical optimizing that has much validity. The absence of good understanding of underlying cognitive questions does mean, however, that we have difficulty in carrying out any sort of speculative or theoretical discussion. The signal failure of physicists and mathematicians adequately to assess or to predict the ways in which computers might be fundamentally useful in teaching is an example of this, and it is

---

It is a paradox in mathematics and physics that we have no good model for the teaching of models.

---

noteworthy how little effect computers have had so far. Indeed, it is a paradox in mathematics and physics that we have no good model for the teaching of models.

This paper will contain, instead, a few informal reflections on the teaching of mathematics and physics at the college level. While the paper will not provide answers to the deeper questions indicated above, it may help the reader to see some of the directions in which answers do *not* lie. Let us begin with a brief taxonomy of introductory mathematics courses. We identify four general categories.

1. *Courses for students who need a credential in mathematics.* Many subject areas require that a student majoring in that area take some college level mathematics. Often, the particular nature of the mathematics is not specified. Calculus, algebra, geometry, logic—all are acceptable. The underlying purposes of such a requirement are to provide experience in quantitative reasoning, practice in working with abstract concepts, and standards of precision and discipline in formal thinking. The hope is that such a course will give the students habits and attitudes of mind that they may then transfer to the conceptual systems of another subject area.

2. *Courses for students in the liberal arts.* These courses occur at two levels. First, there are courses for students who seek a qualitative familiarity with an accessible selection of the ideas that underlie modern mathematics. These students look for some understanding of the essential nature of mathematics and of how mathematics has contributed to the development of contemporary thought and culture. Second, there are courses for students who seek, in addition, directly to engage some of the conceptual, philosophical, and epistemological issues associated with mathematics, even though they do not themselves plan to go on in professional mathematics.

3. *Courses for students who need mathematics for work in other scientific and engineering areas.* These courses also occur at two levels. At the first level, students seek knowledge of mathematical methods that can be directly applied. Such courses are taught partly through drill and partly through providing enough conceptual grounding that the student can recognize situations in which a given method can be applied or for which it requires minor modification. At the second level, such a course puts added emphasis on providing the conceptual equipment with which a student can make a variety of flexible and creative applications. Here the development of a sense of independent mathematical reality becomes especially important. If the student is to construct new mathematical models and then to modify them as circumstances may require, the student must have a strong feeling for the nature and properties of these models independent of the physical reality to which they are being applied.

4. *Courses for students who wish to study mathematics as a professional discipline.* Here, again, courses can occur at two levels. At the first level, the courses are primarily descriptive. They begin to develop the central concepts and methods of modern mathematics and to provide some of the formal details that are necessary for a fully general treatment. At the second level, again as in 2 and 3, a more creative goal is set, and a more active role on the part of the student is expected. The emphasis, from the beginning, is on participation in intellectual activity like that of a research mathematician.

In this paper, we are chiefly concerned on the mathematical side with courses of type 3 at the second level. Such courses, for example, are taken by a majority of students at technological universities. Let us assume that a student is taking a mathematics course of this kind and, at the same time, an introductory course in physics. It is instructive to ask, but difficult to answer, the following question: how do we expect the student to be different at the end of these courses; that is to say, what changes in his mind do we hope will have occurred? While there seems to be no simple and easy answer to this question, we can give some partial and fragmentary answers. We expect the student to have command of a certain collection of mathematical concepts that are used in building models (e.g., vectors, the definite integral, line integrals, flux), command of a certain collection of basic physical concepts and principles (e.g., velocity, work, momentum), and command of a certain collection of computational and conceptual algorithms (e.g., how to differentiate vectors, how to draw conclusions from the law of conservation of momentum). Beyond these, we also expect the student to have developed an intuitive feeling for different ways of understanding and modeling physical phenomena (e.g., microscopic vs. macroscopic, deterministic vs. statistical), some instinctive feeling for the multifarious logic of physical argument (e.g., approaching the same mechanics problem through conservation of momentum, through conservation of energy, or through Newton's second law), and habits of mind and precision of thought upon which new learning and new intuitions can be built.

The thesis of this paper is that for the mathematics teacher and the physics teacher, there is much more to unite them in this enterprise than to divide them. However much the physicist may emphasize physical intuition, students will not get far without a sense of independent mathematical reality. However much the mathematician may emphasize details of the mathematical world, students will lose interest without the sense of power and purpose that physical applications give. Together, the mathematician and the physicist can help to build those most mysterious intellectual achievements within the student's mind, achievements that are central to both professional physics and professional mathematics: the conversion of formal conceptual derivation and representation into confident, quick, and instinctive intuition, and an understanding of how the interplay of guess-work and existing intuition on the one hand and formal derivation and analysis on the other can lead to new (and often surprising) knowledge as well as to new and deeper intuitions.

Here is a brief, specific example of these phenomena. Why is a rotary floor polisher easier to push when it is turned on and rotating than when it is turned off? Most physicists and mathematicians who have not considered the matter before will suggest an incorrect answer, namely, that the coefficient of moving friction of the polishing pad is much lower than the coefficient of stationary friction. In fact, the explanation is both deeper and simpler, and it provides an example in miniature of how an appropriate analysis and model can develop and condition new and useful intuition. In briefest summary, the classical law of friction between two surfaces states that the frictional force on one surface is a vector whose direction is opposite to the motion of that surface and whose magnitude is a constant independent of the motion (and depending only on the normal force of one surface against the other). Applying this to an element of surface at the edge of the polishing pad, and using vector addition of velocities in an appropriate and obvious way, we see that if the pad is rotating and if the polisher is set in motion by a person pushing it, the frictional forces on an element of the pad at three o'clock (as seen by the pusher) will be equal and opposite to the force on an element at nine o'clock (provided that the polisher is not pushed too fast), and that the frictional force on an element at twelve o'clock will be equal in magnitude but not quite opposite in direction to the force on an element at six o'clock. The difference will be the small force of resistance experienced by the pusher. It follows from the geometry that this force, unlike the classical frictional force, must be nearly proportional in magnitude to the velocity of push when the polisher is pushed slowly.

This example (for which I am indebted to a lecture by Professor William Skocpol of Harvard University) has a variety of interesting implications with regard to the conditioning of intuition. On the one hand, a little introspection shows that the usual and mistaken initial suggestion appears to rest on an intuition that the classical frictional force *is* proportional in magnitude to velocity (which we know is wrong when we stop to think about it). On the other hand, we see that our rotating device gives us a situation where the apparent frictional resistance to the pad when we push

---

There are numerous . . . examples in which the interplay of mathematical model and physical circumstance leads to deeper and often surprising intuitions.

---

it truly is proportional to velocity in magnitude. Furthermore, it gives us a good understanding of why the polishing surface is usually fabricated as an annular ring rather than as a full disk. (We leave this as an exercise for the reader.) Finally, it provides us with new intuitive knowledge: we see, for example, why, in pulling a nail from a board or a stake from the ground, it helps to jiggle the nail or stake back and forth while we are pulling (another intuitive exercise for the reader).

There are numerous other examples in which the interplay of mathematical model and physical circumstance leads to deeper and often surprising intuitions. For example, the oscillatory behavior of population sizes in a prey-predator ecological situation is often governed (to a good approximation) by a very simple autonomous system of first-order differential equations. The mathematical study of these equations and of their solution can help to build a new and deeper set of intuitions concerning the balance of an ecological system. A somewhat more complex example, illustrating the interconnected and multifarious nature of physical theory, is the example in relativity of a railroad train that is longer than a tunnel it is to pass through. The train attains a speed sufficiently close to the speed of light so that, from the tunnel-keeper's point of view, it is Lorentz-contracted to be shorter than the tunnel. Hence there is a moment (in the tunnel-keeper's frame of reference) when the train is entirely within the tunnel. At this moment, the keeper drops a massive gate at each end of the tunnel. How is this possible, and what physically occurs to the train in its own initial moving frame of reference? Here, beginning only with the basic Lorentz-transformation formulas, the analysis leads to a vivid intuitive understanding of why rigid bodies cannot exist in relativity theory.

A final, and more sophisticated, illustration of the complex interplay between model and reality occurs in the study of hyperreal numbers. Here, the real numbers themselves, the ultimate ground of the mathematical realities with which the physicist traditionally works, are fundamentally altered in favor of a new and richer mathematical universe where every real number has associated with it a collection of hyperreal numbers that are infinitely close to it on the number line. The intuitive idea of infinitesimal differences is not new, but had been discarded by mathematicians and physicists in the nineteenth century as logically unsatisfactory. The past two decades have seen a logical justification and an elaboration of such an idea; the reader who would like to know more may refer to [1], [2], or [3]. It remains to be seen whether the new models with which this may supply us will lead us, as well, to a new intuitive understanding of physical phenomena.

The above examples suggest some of the subtle but fruitful ways in which models, reality, and intuition can interact in the successful teaching of elementary mathematics and elementary physics. Further pedagogical development in these areas should take account of interactions of this kind. In so doing, it can build stronger foundations for later creative work by students.

References

[1]   K.D. Stroyan and W.A.J. Luxemburg. *Introduction to the Theory of Infinitesimals*. Academic Press, New York, 1976.
[2]   James M. Henle and Eugene M. Kleinberg. *Infinitesmal Calculus*. MIT Press, Cambridge, Massachusetts, 1979.
[3]   H. Jerome Keisler. *Elementary Calculus*. Prindle, Weber & Schmidt, Boston, 1976.

Readin', 'Ritin', and Statistics

Tim Robertson and Robert V. Hogg

Statistics has, for a number of years, been the principle mathematical tool used by scholars in a number of disciplines (e.g., economics, psychology and medicine). For example, a survey of a leading political science journal (see p. xiv of [3]) found that 65 percent of all the articles published between 1968 and 1970 used numerical data. This had grown from 12 percent in the period 1946-48. The uses (and abuses) of statistics are so pervasive that at least one course in statistics is now required by most of the undergraduate programs in the social sciences.

Statistics now plays an important role in our everyday lives. Newspapers are filled with quantitative information measuring, among other things, the popularity of a particular idea or political candidate. The outcome of a presidential election is "determined" after only 10 or 20 percent of the ballots are cast. Television commercials use data to try to convince us that we should purchase a particular brand. If you attend a school board, city council, or faculty meeting it is likely that most of the arguments for one

Tim Robertson is Professor of Statistics at the University of Iowa. He earned his Ph.D. at the University of Missouri in 1966. A Fellow of the American Statistical Association and of the Institute of Mathematical Statistics, Robertson has served on the Council of the American Statistical Association and is currently a Governor of the Mathematical Association of America. He has taught at Cornell College in Mt. Vernon, Iowa and at the University of North Carolina. Robertson is currently an associate editor of the *American Mathematical Monthly*.

Robert V. Hogg is Chairman of the Department of Statistics at the University of Iowa, from where he earned his Ph.D. in 1950. He is a fellow in the Institute of Mathematical Statistics and in the American Statistical Association, and is an elected member of the International Statistical Institute. He has served as a governor of the Mathematical Association of America as well as program secretary for the Institute of Mathematical Statistics for eight years. His publications include over 40 articles from various statistical and mathematical journals and three textbooks, two on statistics and one on finite mathematics. For seven years he was a member of the joint ASA/NCTM committee concerning statistics in the high school curriculum, serving as Chairman for three of those years.

view point or another are supported by summaries of data. For example, at a recent meeting with the principal of a local high school, data was used to convince the audience that: (1) interscholastic sports are not simply for the elite few, (2) girls are not discriminated against in our interscholastic sports programs, (3) football is not an excessively expensive part of our sports programs. In addition we recently heard a committee report on core requirements at the University of Iowa assailed for its lack of hard (numerical) evidence. It is a fact that statistics has become an essential communication skill.

One of the reasons that data are used so frequently to support a certain position is that numerical information is often viewed as being objective and therefore beyond criticism. However, it is easy for numerical information to be very misleading. Consider the claim, made in 1980, that "Ninety-four percent of Chevrolet trucks built since 1970 are still in service." Presumably, we might be interested in this fact because we are a potential purchaser of a truck and because this statement says something about the reliability of Chevrolets. More particularly, it seems to say something about how long we might expect to use one of their trucks before replacing it. In fact, the company seems to want us to believe that if we puchase a Chevrolet truck it is likely to be in service ten years later. Wait a minute! Does this agree with your experience? It doesn't agree with ours. How many ten-year old Chevrolet trucks have you seen recently? If 94% of them were still in service, then the average length of life of Chevrolet trucks must exceed 9.4 years by a substantial amount. Obviously the company's statement is not about ten-year old trucks, but about *all* trucks sold in the period, 1970-79. Suppose a preponderance of these trucks were built since 1977. Those, then, have only been in service for a very few years and it is not surprising that they are still on the road. Obviously the original statement says very little about the probability that a Chevrolet truck will last for ten years, or even for five. Is it objective? We leave it for you to judge.

Consider another example. Suppose a company has 200 employees, 100 of whom are men and 100 of whom are women. Say the average yearly salary of the 100 women is $15,000 while the corresponding figure for the 100 men is $20,000. An obvious case of sex discrimination! But let's be careful and analyze these data further. Suppose we take into consideration the length of employment of these 200 employees, obtaining the following table:

Length of Employment	Men		Women	
	Number	Average Salary	Number	Average Salary
Less than 5 years	20	8,000	80	12,000
At least 5 years	80	23,000	20	27,000

Thus, when length of service is taken into account, it is possible for women to average more than men in each category! (If you do not believe this, you should check our arithmetic for yourself.) Thus these data imply that it is men who in fact are discriminated against in this particular company. (This example was contrived and is not meant to serve as an argument that sex discrimination either exists or doesn't exist. It simply dramatizes the influence of a new variable.) In discrimination situations, as in many problems, it is extremely important to compare individuals with similar characteristics. As this example shows, the subjective choice of which "statistics" to report can change the message. Thus numerical evidence should be consumed not only with a certain amount of skepticism, but also with a considerable degree of skill. All of us should try to be critical consumers of statistics. Doublespeak exists in statistics as much as in the written or spoken word.

As statisticians, we certainly do not want the reader to conclude that statistical analyses are meaningless and should never be accepted to support a particular viewpoint. On the contrary, we are arguing for good statistical analyses. There are great quantities of data being collected today, the summaries of which will influence major decisions. We need bright young (and old) people to make the best possible analyses of these important problems so that reasonably good solutions will evolve. We are talking about such problems as energy, transportation, health, discrimination, education, drugs, pollution, and so on. Moreover, since many of these problems are inter-related, solutions will require decisions about the entire system in which statistics will play a major role. Understanding the subtlety of such systems analysis is vitally important to every informed citizen, as is a critical appreciation of numerical evidence. Statistics has become an extremely important tool, probably ranking in importance behind only the traditional communication skills of reading, writing, speaking, and listening. We believe that at least one course in statistics will eventually be a graduation requirement in many colleges; in fact, one can make a good argument for requiring it in high school.

Teaching statistics

By this point, the reader should recognize that we believe that statistics is not only extremely important in worthwhile research efforts and decision processes, but also in every day living. Good statistical thinking is useful throughout a broad range of activities. The question that now arises concerns statistical education. How do we teach good statistical understanding that will have the proper influence at all levels of daily activities?

Certainly a logical place to start is in the school system, grades K-12, and there has been some effort at that level in this country as well as throughout the world. After some earlier attempts to improve statistical education at this level, Fred Mosteller, as President of the American Statistical Association (ASA), addressed the Annual Meeting of the National Council of Teachers of Mathematics (NCTM) in April of 1967 and called for the creation of a committee to attack the problem of improving statistical education in the schools. This resulted in a Joint ASA/NCTM Committee on the Curriculum in Statistics and Probability, and Mosteller served as the first chairman. Among other things, this committee created two significant publications: *Statistics: A Guide to the Unknown* [4] and *Statistics by Example* [5]. The former is a collection of essays about statistical applications in a number of fields, written for a general audience; the latter is a set of examples that could be used to introduce statistical concepts into existing high school mathematics courses.

The members of this joint committee have spoken at many NCTM meetings supporting the introduction of statistical topics into high school courses. However, the teachers who attended these sessions consistently remark on the lack of text materials and, of course, on the insufficient education of many of them in statistical methods, particularly in "hands on" statistics projects. Even today, as the 1980's begin, these are the two major problems in statistical education at the school level: inadequate material and lack of statistical education among the teachers. Most mathematics teachers can handle the material on elementary probability, but they do have great difficulties with data-oriented courses. The projects in these latter type of courses can motivate students tremendously and it is unfortunate that the teachers do not have the confidence to teach them. Students could collect data around the school or city, concerning what they perceive to be important issues (e.g., drugs, food, scheduling, recreation facilities). Then they could analyse their own data and make recommendations based upon these analyses. They might soon become "experts" in given areas and want their teachers, administrators, and other adults around the community to hear their messages.

Statistics . . . should be used as part of a thoughtful analysis and not as a substitute for it.

While there is really no appropriate textbook for a high school course in statistics (although *Statistics by Example* could supplement certain mathematics courses), there have been some efforts at the college level that might come close to being satisfactory high school texts. However, these would need to be put into a different format for high school students. To appreciate the novelty of these recent textbooks, recall that over the past 30 years, statistics courses offered to satisfy the needs of students majoring in fields such as business, psychology, and education have been essentially "watered down" mathematical statistics courses. The emphasis has been on probability (including combinatorics), on the mathematical properties of

statistical procedures, and on some very specific instructions for drawing conclusions from very special data (usually one sample, single variable data). A typical example asks students to compute a particular statistic, look up a value in a table, and reject or accept a certain hypothesis. These courses say very little about how data should be collected or how statistical analyses can go wrong. Students usually have a difficult time relating the things they learn to their everyday life.

Recently, however, the pendulum seems to be swinging in the other direction. A number of new and exciting books have arrived on the scene; see, for example, [1], [2], and [3]. These books portray statistics as a tool which should be used as part of a thoughtful analysis and not as a substitute for it. They emphasize experimental design, the collection of data, and the interpretation of data. The students are asked not only to be critical consumers of quantitative information, but also to become actively involved in the design, collection, and analysis of data. Such changes are healthy! Indeed, courses of this type should be required of college students and should be greatly encouraged at the high school level. As the variety of textbooks appropriate for such courses expands dramatically in coming years, as we expect that it will, these courses will become both attractive and widely available.

Simultaneously with these developments we must face the issue of appropriate statistical education for teachers. At least one course in statistics must be required in teacher education programs, and it would be highly desirable if two or three could be taken. Assuming that only one course is required, what should be its nature? Certainly not the traditional probability course nor the traditional one in mathematical statistics. It must, on the one hand, include something about elementary design of experiments, collecting data, describing data, models for distributions, and estimating unknown parameters of these models. On the other hand, a good model must fit the data well, and fitting is usually based upon estimation; to understand estimation, substantial consideration must be given to the error structure, including a range of errors arising from the sampling distribution theory all the way to just dumb mistakes. So that all these topics blend together, we envision that in such a 14-week course, the time would be allocated about as follows (but not necessarily in this order):

Collecting, organizing, and describing data	4 weeks
Basic models (some probability and standard distributions)	4 weeks
Statistical inference (mainly estimation with the corresponding error structure)	6 weeks

This course should be designed with a prerequisite of two semesters of calculus and possibly a semester of computer science.

Fortunately, this course would not need to be constructed only for the

small market of prospective high school teachers. We envision that it could
be used to advantage by many mathematics majors, particularly by those
leaning toward the applied mathematical sciences. It is, of course, only a
beginning: we would hope that such a course would be followed by one or
two courses in probability models, mathematical statistics, design and
analysis of experiments, multivariate regression or stochastic processes.

Of course, while we are arguing for more data-oriented courses at the
beginning levels, we recognize the need for strong probability and mathematical statistics courses, along with applied courses, for students entering
Masters and Ph.D. programs. Many interesting and useful developments
have taken place in statistical methods because of theoretical considerations. Advanced students must master basic theory in order to understand
fully the role of statistical methods. However, our plea is that as a student
progresses in advanced programs, he or she must continue to work with real
data, perhaps through such innovative structures as Statistical Consulting
Centers or Mathematical Clinics.

Teachers of these new statistics courses should have broad exposure to
the applications of statistics (such as the experience obtained by working in
a consulting center) in addition to sound background in the fundamentals
of statistics. While most traditionally trained mathematicians can pick up a
textbook in mathematical statistics and teach an acceptable course, *they
cannot teach statistics*. There is a part of statistics that is intrinsically
non-mathematical: there are not always unique solutions in statistical
analyses. Accordingly, we must encourage some of our bright and mathematically talented young people to be applied statisticians by encouraging
them to deal with real data and to try to detect patterns associated with
important problems. This less formalized approach is the nature, not only
of statistics, but much of applied mathematics. We truly hope that others in
the mathematical sciences will see the importance of these investigative
procedures, from kindergarden to the Ph.D. level.

References

[1] David Freedman, Robert Pisani, and Roger Purves. *Statistics*. W.W. Norton,
     New York, 1978.
[2] Dennis G. Haack. *Statistical Literacy: A Guide to Interpretation*. Duxbury
     Press, Massachusetts, 1979.
[3] David S. Moore. *Statistics: Concepts and Controversies*. W.H. Freeman, San
     Francisco, 1979.
[4] Judith M. Tanur, *et al. Statistics: A Guide to the Unknown*. Holden-Day, San
     Francisco, 1972.
[5] Frederick Mosteller, *et al. Statistics by Example*. Addison-Wesley, Massachu-
     setts, 1973.

Mathematization in the Sciences

Maynard Thompson

The superb pictures of Jupiter and Saturn transmitted back to earth by Pioneer 11 impress me as a marvelous technological achievement. Even more impressive is that this was accomplished without several prior attempts. It is certainly the case that subsystems were tested, and that engineers benefited from their experiences with related systems. Nevertheless, the success of this effort in the absence of the usual testing and refinement is remarkable. Contrast this success with the notorious inaccuracy of economic forecasts—despite the best efforts of knowledgeable people and great (dollar) incentives for accurate predictions. We accept success, especially technological success, in the physical sciences and engineering as a matter of course. We are not surprised—disappointed, perhaps, but not surprised—by the lack of success in areas more closely related to the life and social sciences. It is common to say that we understand (or that *someone* understands) the science and engineering of space probes, but we do not have comparable understanding of the economy or of many biological or social systems. What is frequently meant is that there are good mathematical models for the physical sciences, but that the models used in the life and social sciences are not nearly as effective. Let us examine this idea in somewhat more detail.

Maynard Thompson is Professor of Mathematics at Indiana University. He received a B.A. from DePauw University in 1958 and a Ph.D. from the University of Wisconsin in 1962. Thompson has long been concerned with the teaching of applied mathematics from a modelling point of view, serving in 1970-71 as Chairman of the Panel on Applied Mathematics and, in 1973-76, of the Project on Case Studies in Applied Mathematics of the Committee on the Undergraduate Program in Mathematics. His research interests are in biomathematics, especially in population dynamics and epidemiology. Thompson has frequently served as a consultant to academic and business groups and organizations on problems involving mathematical modelling.

Mathematization in the sciences

It is useful to have a term to describe the efforts at systematizing knowledge, especially knowledge of the biological, physical and social worlds, through observation, experiment and study. We use the term *science* in this way. Everyone recognizes biology, chemistry, physics, etc., as sciences. Also, there are areas in economics, psychology, political science, history, and a host of other disciplines which have the characteristics of a science. Certainly the list will be longer ten years from now than it is today. Each science can be measured along a number of dimensions, one of which is its degree of mathematization, that is, the extent to which the ideas and techniques of mathematics are used in the discipline. (Here and elsewhere we use *mathematics* as shorthand for *the mathematical sciences* which include computing, operations research and statistics as well as what is more narrowly referred to as pure and applied mathematics.) From the standpoint of the scientific discipline, a high degree of mathematization need have no special virtue. There are instances in which elaborate mathematical developments have led to no discernible scientific enlightenment just as there are instances in which elaborate equipment has not produced results.

It may be useful to cite a common (but by no means universal) pattern in the ways that mathematics is used in other disciplines. In most cases there has been a tendency toward increased mathematization as a discipline evolves, but this view is conditioned by hindsight and there may be important differences in the future as the role of computers becomes more important. We shall identify four stages which correspond roughly to the degree to which mathematics contributes to the discipline. We emphasize that it is the contribution of the mathematics that is important and not the sophistication or elegance of the mathematics. A relatively simple mathematical idea may have great impact when used ingeniously; on the other hand a very elegant mathematical discussion may contribute little to our knowledge of the scientific problem.

Stage 1. Mathematics frequently enters into scientific work with the collection, organization and interpretation of data and information. In some instances there are large amounts of data collected from observations: Tycho Brahé's astronomical data, Gregor Mendel's data on plant reproduction, and the results of the numerous experiments in psychological learning beginning in the 1930's. In other instances, such as relatively uncommon natural phenomena (major earthquakes, rare diseases, etc.), there may be relatively few instances to examine. The task of deciding which data are needed and how to go about the collection process is frequently a difficult one. The task is, however, essential and the absence of good data can significantly retard the general development of an area, and its mathematization in particular [7]. For instance, work on the question of how information (or rumors) spread through social organizations has progressed slowly due in part to the paucity of good data. From the view of interaction with mathematics, data collection and analysis are the domain of the statistician and, at times, the computer scientist. Much mathematical research in statistics and computing has been stimulated by challenging questions arising in the treatment of real data.

Stage 2. Some collections of data display only the most general regularities—economic statistics frequently are of this sort. Other collections display remarkable patterns when examined closely. Two of the best examples of this are Brahé's data on the motion of the planets and Mendel's records of plant hybridization experiments. Sufficiently regular patterns in data or information may be summarized in empirical "laws of nature." These laws, which at this point are no more than convenient ways of summarizing observations, may be very precise or quite vague. In the physical sciences such laws are frequently quite precise. For instance, Kepler's laws of planetary motion, which are empirical laws deduced from Brahé's observations, and Snell's law of refraction are very precise statements. Empirical laws tend to be much less precise in the life and social sciences. For instance, Gause's law of competitive exclusion (two different species competing for exactly the same biological niche cannot coexist) and Gossen's law of marginal utility (the marginal utility of a good decreases as more of that good is consumed) are much less precise and their legitimacy can be debated in a way that Snell's law cannot be.

There are some caveats worth mentioning in discussions of empirical laws. First, such laws are frequently derived on the basis of aggregate behavior: since the data are smoothed or averaged, the law may be more applicable to a (perhaps nonexistent) "average" situation than to any specific situation. Second, there are frequently (implicit or explicit) assumptions underlying the statement, and the law may describe observations only to the extent that the assumptions are fulfilled. For instance, one of Mendel's laws of genetics asserts that genes relating to different characteristics recombine at random. This law provides an accurate description only of the behavior of genes which are not linked. If we are interested in studying characteristics associated with linked genes, then this law does not apply.

Empirical laws which serve as succinct and illuminating summaries of experiment and observation may be one of the goals of an experimentally inclined scientist. Such laws may be midway in the analysis of a situation by a theoretically oriented scientist, and they may be near the beginning of the involvement of a mathematical scientist.

Stage 3. Following the identification of empirical laws, the next stage in the mathematization of a science is the creation of a mathematical structure —sometimes referred to as a theory—which accounts for these laws. The expression "accounts for" requires elaboration and we shall return to it later. For the moment, the usual intuitive meaning will suffice. In both senses, intuitive and precise, the meaning will vary from one discipline to another and from one time period to another. A mathematical structure together with an agreement as to the relations between the symbols and notation of the mathematics and the objects and actions of a real world situation is known as a mathematical model of that situation. There may be more than one mathematical structure which accounts for an empirical law, or there may be one structure which accounts for some of the laws of a situation and another structure which accounts for other laws. An example of the latter from elementary physics is the dual wave and particle models for light: the wave model provides explanations for the phenomena of physical optics (reflection, refraction, dispersion), while the photo-electric effect—which poses problems in a wave model—is perfectly comprehensible in a particle model.

An investigation of the mathematical structure leads to conclusions, usually called theorems, which are deduced by logical arguments from the assumptions and definitions. Using the agreements between the symbols and terms in the mathematics and the original setting, these theorems can be translated into assertions, usually called predictions, about the real world setting. We can now sharpen our meaning of "accounts for." We say that a mathematical model accounts for an empirical law if one of the theorems of the model translates into that law.

The task of determining whether a model accounts for a set of observations is frequently a very complex one. It is common for models to involve parameters, symbols which represent real world quantities such as the velocity of a particle, the cost of a commodity, the arrival rate of cars at a tollbooth, or the likelihood that a subject will remember a nonsense word after a single presentation. In order to compare predictions with observations one must know the values of the parameters which appear in the model. The estimation of parameters is a challenging mathematical and scientific task, and many otherwise apparently acceptable models have not been pursued because they relied on parameters which could not be estimated.

A mathematical model, as a mathematical structure, may of course raise questions which are interesting from a mathematical standpoint. The questions may, but need not, have significance for the situation which gave rise to the model. However, one of the great advantages of phrasing things in mathematical form is that the same or very similar structures may arise in quite different scientific settings. Questions which are of little or no interest in one setting may be highly important in another setting. Thus, the development of a model beyond immediate needs is quite common.

The merit of a mathematical model viewed as a scientific contribution to the study of some phenomena is determined by the degree to which the predictions of the model agree with observations. Normally model building is a cyclic process and the first attempt does not yield predictions which agree closely with observation. In such a case the model must be modified and a new set of predictions derived. It is common for a model to be refined several times before adequate agreement is obtained, and this refinement may continue over many years.

The role of the mathematician in the modelling process is that of a collaborator who sometimes makes major contributions but who rarely sets the broad directions for the study. None of the activities described in stages 1 and 2 is purely mathematical, and of that described in stage 3, only the analysis of the mathematical structure does not also include science. Even

---

The creation ... validation or interpretation of ... a model is normally possible only by someone very familiar with the science of the situation.

---

there scientific insight may be helpful. The creation of a model and the validation or interpretation of predictions based on the model are normally possible only by someone very familiar with the science of the situation.

Stage 4. Suppose that we have created and developed a mathematical model, and derived a body of predictions (theorems) which account for some observations. What more can be done? We can examine the model with a goal of obtaining *new* scientific insight. That is, we look for aspects of the science of the situation that are either uncovered or illuminated by the use of the mathematical model. Frequently this means that the mathematics leads to predictions of scientific phenomena which are subsequently observed. In stage 3 we used existing data to validate the model, here we use the model to tell us where to look for data.

Modern physics offers many examples. (The Messenger Lectures of Richard Feynman [3] provide a fine overview.) As an instance we cite the Weinberg-Salem Theory which predicted right and left handed electrons (a right handed electron has its spin oriented in the direction of its motion just as an ordinary screw moves forward when turned to the right) and the subsequent observation of them in experiments at the Stanford Laboratory. The observations fit very closely with the predictions. Other good examples occur in chemical kinetics. In particular, a type of chemical reaction known as the Belousov-Zhabotinskii reaction exhibits regular sustained oscillations between states (colors), and in some cases outward propagating circular waves. An analysis of a mathematical model for the situation resulted in a prediction of inward propagating waves as well. Waves of this type have been observed after their existence was predicted and an effort was made to find them. A confirmation such as this greatly enhances our confidence in using a model as a description of the real world setting.

To a theoretically oriented scientist, the use of mathematics to gain new insight into science may be the most rewarding goal. Others may have different objectives; for instance, the experimentalist may find the theoretical support of observations most satisfying while a mathematician may find a new mathematical problem most exciting.

Having discussed the four stages in some detail, we can now summarize them with a few key phrases:

Stage 1: Data and information collection, analysis and interpretation.
Stage 2: Quantitative formulation of scientific principles and empirical laws.
Stage 3: Formulation, study and validation of mathematical models.
Stage 4: Use of mathematical models to gain scientific insight.

Most situations which arise in the physical sciences and engineering are approached in terms of stage 3 or stage 4. Some areas in biology, especially genetics, population biology and ecology have models in stage 3. A few areas in psychology, primarily learning and memory theory, also have stage

---

Models which contribute to new scientific insight are relatively uncommon in the life and social sciences.

---

3 models. Models which contribute to new scientific insight are relatively uncommon in the life and social sciences but there are some examples. For instance, models in population dynamics involving the mathematical phenomenon of "chaotic behavior" can be viewed as providing scientific insight, although the biological implications of the model are by no means completely understood [4]. This surprising result is that a system whose behavior through time is completely determined step by step (by a simple equation) can, for certain values of parameters, display global behavior which appears totally random. Another instance is given by the work of Kenneth Arrow on the relations between individual and group preferences. (Arrow's original work is [1]. The connections between Arrow's ideas and more general decision problems are described in [2].) His goal was to study group decision methods with certain intuitively desirable properties. The situation was dramatically illuminated by his recognition and proof of the inconsistency of a natural axiom system (that is, a mathematical model) for group decision making. Simply stated, there is no group decision making process which reflects individual preferences in these natural ways.

However, a substantial majority of the work in the social and life sciences is in stages 1 and 2. There is, moreover, some divergence of opinion as to the extent to which mathematics, which has proved so effective as a language in the physical sciences, will be equally effective as a language in other sciences. The prospective user of quantitative techniques in the life and social sciences should be sensitive to the potential hazards as well as the potential rewards of the activity [6].

Goals of mathematization

An experimentalist may organize and interpret results simply to make the mass of information accessible to others. Likewise, empirical laws may be formulated primarily to communicate in abbreviated form the results of many observations. Frequently the formulation of empirical laws may be accompanied by a sharpening of the concepts and the language used to describe them. Thus, facilitating communication and sharpening concepts are common goals.

Two other goals of the use of mathematical models that deserve special attention are to gain understanding of the scientific situation and to aid in making decisions. The idea of using models to gain understanding was introduced above, and we shall return to it in more detail in a moment. First, let us turn to the economically important use of models in making decisions. Certainly, many investigations will involve several goals and to some extent all models have a goal of understanding. Nevertheless, models for making decisions have some distinguishing features.

A typical situation is one in which a planner or manager must select among a number of alternative courses of action with an objective of optimizing some quantity (maximizing profits or minimizing costs, for instance). It may be that the environment in which the decision is to be made involves some unknown factors (demand for a product, for example) and perhaps the actions of a competitor. There are well developed models utilizing mathematical programming and game theory which may be helpful in selecting among various alternatives. The concept of human rationality and the way it is reflected in actions plays a role in many of these models. It is interesting that there is evidence that knowledgeable individuals who have not explicitly posed their problems in mathematical terms nevertheless act in ways which are consistent with the actions suggested by a mathematical analysis. (The very readable book [5] discusses several major business decisions which can be viewed in a game theory model.)

When discussing the use of mathematics to gain understanding of scientific situations there are several things we could mean. First (and most obvious), we could mean that there are relations between aspects of real situations that we recognize after studying a mathematical model that we did not recognize before. For instance, an identification of the role of core infectives in the spread of gonorrhea helps in understanding the response of the disease to various forms of treatment. Another example is the understanding of the strategic significance of different voting procedures which results from a study of qualitative voting models [2]. Finally the work of Kenneth Arrow mentioned previously provides a third example. Arrow's work illustrates very well the contribution that can be made by precise formulation and careful analysis of the consequences of different assumptions. An analysis of the precise model he constructed provides much more detailed information concerning the influence of assumptions on conclusions than can be obtained otherwise. Before the consequences of an assumption can be investigated, the assumption itself must be made precise. This is part of the modelling activity.

There is another possible meaning for the understanding gained from the study of a mathematical model. Since most models include parameters, and the predictions depend on these parameters, we gain knowledge of the situation if we can determine how or to what extent the predictions change with variations in the parameters. This knowledge can be both practically and theoretically useful. As an illustration, consider an economic allocation problem in which some quantity, say profit, depends on various inputs which are subject to certain constraints. The goal is to determine a permissible selection of inputs which maximizes the profit. Suppose that a specific input is constrained so that no more than amount A is available. One might be interested in the way maximum profit changes as A varies. (This change is an economically important quantity known as the shadow price of the input.)

Finally, understanding may mean knowing where to look next for new results. The model may suggest interesting new experiments, new ways of analyzing or interpreting existing data, or entirely new conceptual formulations. For instance, some ecological models have raised questions regarding the interactions of populations which suggested new experiments.

Most of the mathematical ideas and techniques currently being used in all fields are the direct descendents of those developed for studying problems in the physical sciences. In general, the use of mathematics begins at stage 1 and progresses as more and better data become available and we learn more about the situation. The fact that most of the life and social sciences are currently using mathematics in stage 1 or 2 is not unexpected. If the pattern set by the physical sciences holds in these areas, we can expect important new scientific insights to result from the continued mathematization of these sciences.

References

[1] K.J. Arrow. *Social Choice and Individual Values*. Cowles Foundation Monograph, No. 12, Wiley, New York, 1963.
[2] S.J. Brams. *Game Theory and Politics*. Macmillan, The Free Press, New York, 1975.
[3] R. Feynman *The Character of Physical Law*. MIT Press, Cambridge, 1967.
[4] R.M. May. "Biological Populations with Nonoverlapping Generations: Stable Points, Stable Cycles and Chaos." *Science* 186 (1974) 645-647.
[5] J. McDonald. *The Game of Business*. Doubleday, New York, 1975.
[6] A. Rapoport. "Mathematical Methods in Theories of International Relations: Expectations, Caveats, and Opportunities." In *Mathematical Models in International Relations*. Ed. by D.A. Zinnes and J.V. Gillespie. Praeger, New York, 1976.
[7] H.A. Simon. "The Behavioral and Social Sciences." *Science* 209 (1980) 72-78.

STRANMILLIS COLLEGE
TRENCH HOUSE
BELFAST BT11 9GA